EVOLVING BRAINS, EMERGING GODS

EVOLVING BRAINS, EMERGING GODS

EARLY HUMANS *and the* ORIGINS OF RELIGION

E. FULLER TORREY

Columbia University Press
New York

Columbia University Press
Publishers Since 1893
New York Chichester, West Sussex
cup.columbia.edu

Library of Congress Cataloging-in-Publication Data
Names: Torrey, E. Fuller (Edwin Fuller), 1937– author.
Title: Evolving brains, emerging gods: early humans and the origins of religion / E. Fuller Torrey.
Description: New York, New York: Columbia University Press, [2017] | Includes bibliographical references and index.
Identifiers: LCCN 2017010259 ISBN 9780231183369 (cloth: acid-free paper) | ISBN 9780231544863 (e-book)
Subjects: LCSH: Religion—Philosophy. | Anthropology of religion. | Brain—Evolution.
Classification: LCC BL51.T6155 2017 | DDC 200.1—dc23
LC record available at https://lccn.loc.gov/2017010259

∞

Columbia University Press books are printed on permanent and durable acid-free paper.
Printed in the United States of America

Cover design: Milenda Nan Ok Lee
Cover image: DEA / G. Nimatallah © Getty Images

FOR BARBARA,
WITHOUT WHOM THE BOOK WOULD
NOT HAVE BEEN WRITTEN,
WITH THANKS FOR FIFTY
GREAT YEARS

I have wrestled with death. It is the most unexciting contest you can imagine. It takes place in an impalpable greyness, with nothing underfoot, with nothing around, without spectators, without clamour, without glory, without the great desire of victory, without the great fear of defeat, in a sickly atmosphere of tepid skepticism, without much belief in your own right, and still less in that of your adversary.

—JOSEPH CONRAD, *HEART OF DARKNESS*, 1903

Civilizations, economic systems, migrations, war and peace may come and go; but the question of death insistently remains. And it links together in one common humanity—perplexed and distressed—all the thousand upon thousand generations of men, all the myriad tribes, races and nations, all the varying groups, types and classes of mankind.

—CORLISS LAMONT, *THE ILLUSION OF IMMORTALITY*, 1935

CONTENTS

Preface *xiii*
Acknowledgments *xvii*

INTRODUCTION: THE BRAIN, HOME OF THE GODS 1

An Evolutionary Theory 3
The Human Brain 6
The Nature of the Evidence 9
Parallel Evolution 16

PART 1 THE MAKING OF THE GODS

1 *HOMO HABILIS*: A SMARTER SELF 23

The First Hominins 25
The Brain of *Homo habilis* 29
Basic Areas Associated with Intelligence 31
Why Did the Brain Increase in Size? 35

2 *HOMO ERECTUS*: AN AWARE SELF 37

Self-Awareness 40
The Brain of *Homo erectus* 45
A Self-Aware Neuron? 48

3 ARCHAIC *HOMO SAPIENS* (NEANDERTALS): AN EMPATHIC SELF 51

A Theory of Mind 55
Do Animals Have a Theory of Mind? 57
When a Theory of Mind Is Impaired 59
The Brain of Archaic *Homo sapiens* 61
Theory of Mind and Belief in Gods 65

4 EARLY *HOMO SAPIENS*: AN INTROSPECTIVE SELF 68

The First Sparks 69
An Introspective Self 74
The Introspective Self and Language 78
The Introspective Self and the Gods 82
The Brain of Early *Homo sapiens* 84

5 MODERN *HOMO SAPIENS*: A TEMPORAL SELF 87

Intentional Burials with Grave Goods 92
The Advent of the Arts 95
Mastering the Future: The Evolution of Autobiographical Memory 104
The Emergence of Religious Thought 1: The Meaning of Death 110
The Emergence of Religious Thought 2: The Meaning of Dreams 117
The Human Revolution Revisited 120
The Brain of Modern *Homo sapiens* 125

PART 2 THE EMERGENCE OF THE GODS

6 ANCESTORS AND AGRICULTURE: A SPIRITUAL SELF 133

"The First Human-Built Holy Place" 134
Ancestor Worship 137
The Domestication of Plants and Animals 140
Farming and Parallel Evolution 144
The Living and the Dead 146
Skull Cults 149
The Earliest Gods 155
The Brain of the First Farmers 161

7 GOVERNMENTS AND GODS: A THEISTIC SELF 165

Mesopotamia: The First Documented Gods 166
Gods in Other Early Civilizations 176
The Emergence of Major Religions 195

8 OTHER THEORIES OF THE ORIGINS OF GODS 203

Social Theories 206
Prosocial Behavior Theories 208
Psychological and Comfort Theories 211
Pattern-Seeking Theories 213
Neurological Theories 215
Genetic Theories 217
Are Gods the Products, or By-Products, of Evolution? 218

Appendix A: The Evolution of the Brain 225
Appendix B: Dreams as Proof of the Existence of a Spirit World and Land of the Dead 229
Notes 239
Index 279

PREFACE

I have been looking for God, indeed any god, since I was a boy. As an acolyte in my local church, I assisted in serving communion and was told that God was there. As a university student, I majored in religion and studied various manifestations of the gods. As a graduate student in anthropology, I discovered surprisingly similar gods in very dissimilar cultures. As a physician and psychiatrist, I have studied the brain and wondered where in it the gods might reside. Seventeenth-century British physician Thomas Willis, the first person to systematically study the brain, was correct in saying that such studies "unlock the secret places of man's mind." I have also visited many of the world's shrines built to honor gods and have inhaled the numinous ether that pervades them all. I am especially fond of Gothic cathedrals, which, when filled with choral music, may become transcendent.[1]

It was while visiting one of these shrines, at Avebury in England, that I decided to write this book. Sitting on the terrace of the Red Lion Pub, I could see 4,500-year-old Silbury Hill, 130 feet in height, the tallest man-made earthen mound in Europe. It had been built with remarkable engineering ingenuity, using bone and wooden tools, in a series of radial compartments, so that even today there are few signs of erosion. At the same time that Silbury Hill was being built, Egyptians at Saqqara were building the first stepped pyramid, 200 feet in height; Peruvians at Caral were building a platform mound, 100 feet in height; and the Chinese at Chenzishan were building a massive platform with a temple on top. Earthen mounds and pyramids would subsequently be built in many other parts of the world, including Indonesia, Sudan, Mexico, Guatemala, Honduras, and the United States, such as the 100-foot-high Monk's Mound at Cahokia, near St. Louis. These were all probably built to reach and honor the gods, a logical response to human needs arising from the evolution of our brain.

However, it is important to keep in mind that our present theories about gods are based on incomplete information. We still have much to learn about how the human brain evolved and how it functions. Our knowledge of the evolution of *Homo sapiens* and the development of religious ideas is also fragmentary. Many of the most important archeological finds have been discovered accidentally. For example, the 28,000-year-old burials at Sungir, Russia, were found while removing clay from a pit; similarly, the extraordinary finds at Varna in Bulgaria, Ain Ghazal in Jordan, Nevali Çori in Turkey, Wuhan in China, and Garagay in Peru were all accidently uncovered during construction projects, while Göbekli Tepe in Turkey and the Ness of Brodgar in Scotland were uncovered by farmers plowing their fields. There are presumably hundreds of similar sites yet to be discovered; they should provide us with additional details regarding the evolution of *Homo sapiens* and the emergence of gods. What follows is thus provisional, based on the facts as presently known.

In describing human evolution, I have generally avoided geological and archeological period terms and instead used a continuous measure of years before the present. When precise dates were needed, I have used BCE (before the common era) and CE (common era). I have also used contemporary geographical names for most ancient places to help readers identify the locations. In keeping with modern terminology, I use *hominids* to refer to all great apes, including humans, and *hominins* to refer to the human line, including *Homo sapiens* and all our immediate ancestors, that separated from the great apes about six million years ago. To assist readers, I have placed the detailed brain information in clearly marked sections for those who wish to skip it, and I have grouped references together at the end of the paragraph.[2]

The terms *gods* and *religion* are both problematic, because they have been used so variably by different scholars. Some have argued that anything that has supernatural powers is a god, including ancestor, animal, and nature spirits. I am using *gods* in a more restricted sense to indicate male or female divine beings who are immortal and who have some special powers over human lives and nature. Even this definition covers a wide range of gods with varying degrees of omniscience, omnipotence, and omnipresence, who may or may not have created the earth and humans, and who may or may not be concerned with human events. Gods who are completely divorced from all human events are sometimes referred to as *high gods*. When *God* is capitalized, it refers to the monotheistic deity of Judaism, Christianity, and Islam. *Religion* is also a very broad and imprecise term used to refer to everything from a feeling of spirituality to a set of beliefs and rituals. This book will not attempt to provide a precise definition of *religion* but rather demonstrate how the emergence of gods led to the development of religion in its many manifestations. When I use the term, I am using it to refer to "the feelings, acts, and experiences of individual men . . . in relation to whatever they may consider the divine," with "divine" meaning "godlike," as defined by William James.[3]

The evolutionary journey of *Homo sapiens* that brought us gods and formal religions has been truly extraordinary. Not only did our brain evolve, but it also evolved in a way that enables us to comprehend the process by which it evolved, to write about the process, and to think about its implications for our lives.

ACKNOWLEDGMENTS

My largest debt is to Wendy Lochner, the editor at Columbia University Press who believed in the manuscript despite the fact that it mixes multiple disciplines and defies easy categorization. Carolyn Wazer, Lisa Hamm, Robert Demke, and all who assisted with publication were highly professional and a pleasure to work with. I am also grateful to Maree Webster, who generously provided neuroanatomical expertise for the brain drawings. Similarly, I owe many thanks to Andrew Dwork and Jeffrey Lieberman for correcting my many neuroanatomical misconceptions.

Many people patiently responded to my queries. They include Christiane Cunnar at the invaluable Human Relations Area Files at Yale University, Tim Behrens, Todd Preuss, Tom Schoenemann, and Sara Walker. Many others read sections of the manuscript in its various stages; I want to especially thank Halsey Beemer, John Davis, Faith Dickerson, Jonathan Miller, Robert Sapolsky, Robert Taylor, Maynard Toll, and Sid Wolfe. I thank Faber and Faber for permission to quote from T. S. Eliot's *Four Quartets*. Finally I want to express my gratitude to my research assistants, Judy Miller and Wendy Simmons, and to Shakira Butler and Shen Zhong, who provided administrative support.

EVOLVING BRAINS,
EMERGING GODS

INTRODUCTION

The Brain, Home of the Gods

It is essential to understand our brains in some detail if we are to assess correctly our place in this vast and complicated universe we see all around us.

—FRANCIS CRICK, *WHAT MAD PURSUIT*, 1988

Where did the gods come from? And when did they come? These questions were the impetus for the writing of this book. Psychoanalyst Carl Jung claimed that "all ages before ours believed in gods in some form or another." But is this necessarily true? Did ancient hominins also have gods? By contrast, religion researcher Patrick McNamara argued that the existence of gods and their attendant religions is one of the most distinctive characteristics differentiating modern *Homo sapiens* from our hominin forefathers—"as emblematic of its bearer as the web for the spider, the dam for the beaver, and the song for the bird."[1]

Wherever and whenever they came, it is clear that believing in one or more gods is a deeply felt human need. A poll of Americans conducted in 2012 reported that 91 percent said they believe in God or a "Universal Spirit," with three-quarters saying they are "absolutely certain" such a deity exists. Such belief supports Jean-Jacques Rousseau's description of humans as "theotropic creatures, yearning to connect our mundane lives, in some way, to the beyond." Indeed, our desire for the divine is so strong that Francis Collins, an eminent scientist and devout Christian, has argued that the "universal longing for God" is itself proof of the existence of a purposeful Divine Creator. Almost three thousand years ago, Homer similarly noted that "all men need the gods."[2]

Judaism, Christianity, and Islam teach that there is one God, but most religions claim there are many. Indeed, there is an abecedarian abundance of them, from Ahura Mazda, Biema, Chwezi, Dakgipa, Enuunap, Fundongthing, Great Spirit, Hokshi Tagob, Ijwala, Jehovah, Kah- shu-goon-yah, Lata, Mbori, Nkai, Osunduw, Pab Dummat, Quetzalcoatl, Ra, Sengalang Burong, Tirawa, Ugatame, and Vodu, to Wiraqocha, Xi-He, Yurupari, and Zeus. Sixteenth-century French essayist Michel de Montaigne noted man's propensity for making gods when he wrote: "Man is certainly stark mad. He cannot make a worm and yet he creates gods by the dozens."[3]

The gods are also ubiquitous, found everywhere on earth, in the heavens, and in regions under the ground. Some gods have been associated with particular places, such as Athena with Athens; others with forces of nature, such as Poseidon with the sea; and still others with human endeavors, such as Aphrodite with love. In monotheistic religions, a single god is often responsible for all human activities, while in polytheistic religions there may be an extraordinary degree of divine specialization. In ancient Rome, for example, three different gods (Vervecator, Reparator, and Imporcitor) were associated with the three times that fields were plowed; another god (Insitor) with sowing the seeds; another (Sterculinius) with spreading the manure; another (Sarritor) with weeding the field; another (Messor) with reaping the grain; and still another (Conditor) with storing the grain. Perhaps the ultimate expression of divine specialization was achieved by Polynesians in Tonga who had "a special god assisting thieves in their trade." Throughout human history, new gods have appeared and old gods have died. Live gods are found in places of worship, whereas many dead gods are found in museums, where their images are regarded as works of art.[4]

AN EVOLUTIONARY THEORY

Concerning where the gods came from, this book will argue that they came from the human brain. In regard to when they came, the book will argue that the gods arrived after the brain had undergone five specific cognitive developments. Such developments were necessary for being able to conceive of the gods. As Homo habilis, about 2 million years ago, hominins experienced a significant increase in brain size and general intelligence (chapter 1). As Homo erectus, beginning about 1.8 million years ago, they developed an awareness of self (chapter 2). As Archaic Homo sapiens, beginning about 200,000 years ago, they developed an awareness of others' thoughts, commonly referred to as having a "theory of mind" (chapter 3). As early Homo sapiens, beginning about 100,000 years ago, they developed an introspective ability to reflect on their own thoughts. Thus, they could not only think about what others were thinking but also about what others were thinking about them and their reaction to such thoughts (chapter 4).

Finally, as modern *Homo sapiens*, beginning about 40,000 years ago, we developed what is commonly referred to as an "autobiographical memory," an ability to project ourselves backward and forward in time. We were thus able to predict and more skillfully plan for the future. For the first time in hominin history, we could therefore fully understand death as the termination of our personal existence. And for the first time, we could envision alternatives to death, including places where our deceased ancestors may still exist (chapter 5).

Arguing that a specific cognitive skill is associated with a specific stage of hominin evolution of course does not mean that this skill developed only at that time. All cognitive skills evolved as part of the entire course of hominin evolution and presumably are continuing to evolve. Associating a specific cognitive skill with a specific stage of hominin evolution simply means that, at that stage of evolution, hominins exhibited some new behavior of which we are aware, and this behavior suggests that this particular cognitive skill had matured to the point of being able to affect the behavior of these hominins. For example, approximately 100,000 years ago, shells that apparently were used to make decorative necklaces first appeared. This suggests that the cognitive ability of hominins to think about what other hominins were thinking about them had matured to the point where it was affecting their behavior. Precursors of this cognitive skill may have existed 100,000 years earlier, and it may have become even better developed 50,000 years later, but we regard the shells used as decoration as a marker of cognitive evolution.

The acquisition of an autobiographical memory and other cognitive skills led to the agricultural revolution, beginning about 12,000 years ago. This brought people together to settle in villages and towns for the first time and produced a dramatic increase in the population.

Living in one place allowed the dead to be buried next to the living; consequently, ancestor worship became increasingly important and elaborate. As populations increased, hierarchies of the ancestors inevitably emerged. At some point, probably between 10,000 and 7,000 years ago, a few very important ancestors crossed an invisible line and conceptually became regarded as gods (chapter 6).

By 6,500 years ago, when the first written records became available, gods had become numerous. Initially, their responsibilities focused on sacred issues of life and death. However, political leaders soon recognized the usefulness of the gods and increasingly assigned them secular duties as well, such as administering justice and waging war. By 2,500 years ago, religion and politics were supporting each other, as the major religions and civilizations became organized (chapter 7). In the final chapter, the utility of a brain evolution theory of the gods will be compared with other theories that have been proposed (chapter 8). The utility of any theory should be assessed by its ability to explain the known facts.

The evolutionary theory of gods proposed in this book is not original. Rather, it is an updated version of a theory first proposed, fittingly, by the father of evolutionary theory, Charles Darwin. As a young man,

Darwin had held traditional Christian beliefs and had even considered entering the ministry. During his five-year voyage on the *Beagle*, he later recalled, he had been "heartily laughed at by several of the officers . . . for quoting the Bible." When Darwin returned to England and started developing his theory of natural selection, it occurred to him that religious belief might also be a consequence of brain evolution. In his personal notebook Darwin wrote that he had "thought much about religion," and in his typical telegraphic writing style, he speculated that "thought (or desires more properly) being hereditary" might be "a secretion of the brain." If this were true, he continued, "it is difficult to imagine it [belief in God] anything but structure of brain hereditary . . . love of deity effect of organization." Thus, thoughts, desires, and "love of deity," he speculated, were all products of our brain organization.[5]

Darwin, only twenty-nine at the time, was not about to express such thoughts publicly. He was aware that his emerging theories of natural selection were sharply at variance with the Christian belief that man had been made in the image of God; his reluctance to offend the religious establishment as well as his pious wife was a major reason why he did not publish his theories of natural selection for another twenty years.

Just as Darwin's views on natural selection were shaped by the animals he had encountered on his voyage around the world, so his views on gods were shaped by the people he had encountered. He had met native people in South America, in New Zealand, in Australia, in Tasmania, and on myriad islands across the Atlantic and Pacific Oceans and had been impressed by their many gods. In The Descent of Man, he noted that "a belief in all-pervading spiritual agencies seems to be universal," and this "belief in spiritual agencies would easily pass into the belief in the existence of one or more gods." Foreshadowing theories of brain development, Darwin added that such beliefs only occur after a "considerable advance in the reasoning powers of man, and from a still greater advance in his faculties of imagination, curiosity, and wonder." Darwin likened "the feeling of religious devotion" in humans to "the deep love of a dog for his master" and cited a writer who claimed that "a dog looks on his master as on a god."[6]

In later years, Darwin's theories led him to a complete disbelief in God. In his autobiography, he wrote: "Disbelief crept over me at a very slow rate but was at last complete. The rate was so slow that I felt no distress, and have never since doubted even for a single second that my conclusion was correct." As with many people, the problem of evil contributed to Darwin's ultimate loss of faith. He was especially troubled by the death of his favorite daughter at age ten from what was probably tuberculosis. Darwin also asked how a supposedly omnipotent and omniscient God could allow "the sufferings of millions of the lower animals throughout almost endless time." To a friend he wrote: "I cannot see, as plainly as others do, evidence of design and beneficence on all sides of us. There seems to me too much misery in the world." Ultimately, Darwin even failed to perceive a deity in the process of creation, finding "no more design in the variability of organic beings and in the action of natural selection, than in the course which way the wind blows."[7]

THE HUMAN BRAIN

In order to assess an evolutionary theory for the emergence of gods, it is necessary to understand something about the human brain. This will be briefly summarized in this chapter, with more details provided in the notes and appendixes. The brain is a wondrous organ, thought to comprise 100 billion neurons and 1,000 billion glial cells. If you decide to give away your brain cells, there will be enough to give 16 neurons and 160 glial cells to every person on earth. Each neuron is connected to at least 500 other neurons, resulting in a total of 100,000 miles of nerve fibers in each brain; if laid end to end, these nerve fibers could circle the earth 4 times. The nerve fibers are covered with myelin, a light-colored substance; because it is light in color, the nerve fiber connecting tracts are referred to as "white matter." The neurons, glial cells, and connecting tracts together create infinitely complex brain networks, making the human brain the most complex object known in the universe. British

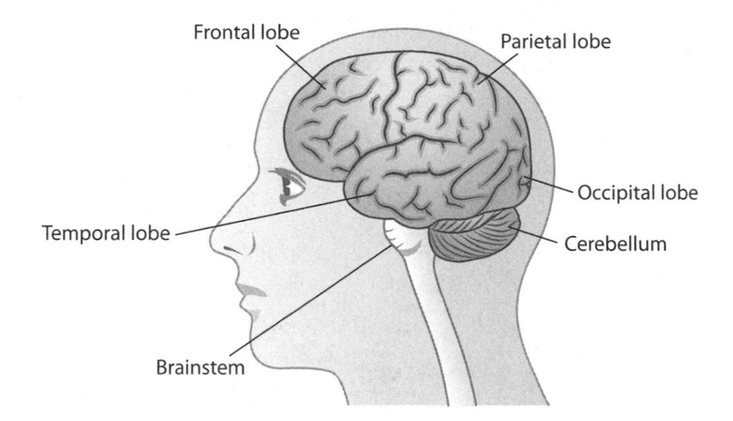

FIGURE 0.1 The four lobes of the brain.

neurologist Macdonald Critchley described it as "the divine banquet of the brain . . . a feast with dishes that remain elusive in their blending, and with sauces whose ingredients are even now a secret."[8]

Topographically, the human brain is divided into two hemispheres, each of which has four major lobes: frontal, temporal, parietal, and occipital (figure 0.1). It is further subdivided into 52 separate areas based on the organization of brain cells as seen under a microscope. The original division of brain areas was done in 1909 by Korbinian Brodmann, a German anatomist; it has been modified several times over the years, but the brain areas are still referred to as Brodmann areas, usually abbreviated BA plus a number, for example, BA 4. The Brodmann numbering system will be used in this book for readers interested in the localization of brain functions. Figure 0.2 shows the Brodmann areas.[9]

Neuroimaging studies and postmortem brain studies have shown which human brain areas evolved first and which evolved more recently, as detailed in appendix A. The brain areas that evolved most recently are often referred to as "terminal areas," so named by Paul Emil Flechsig, a German researcher. Importantly, these most recently evolved brain areas are the same areas that are associated with most of the cognitive skills that make us uniquely human. Neuroimaging studies have also determined the order in which the white matter tracts, which connect

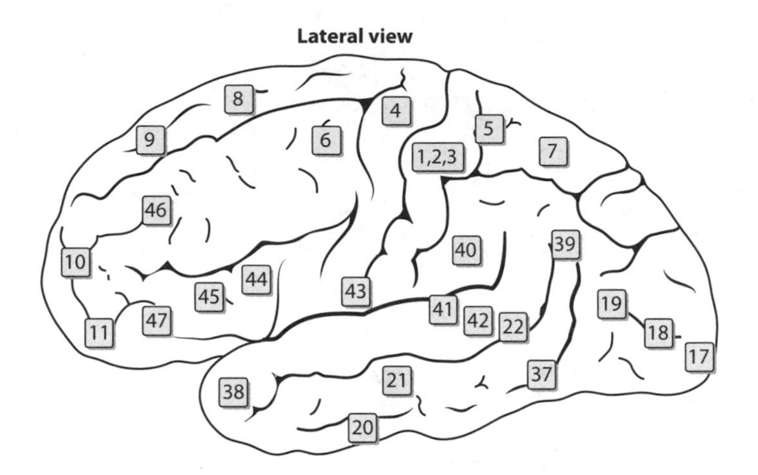

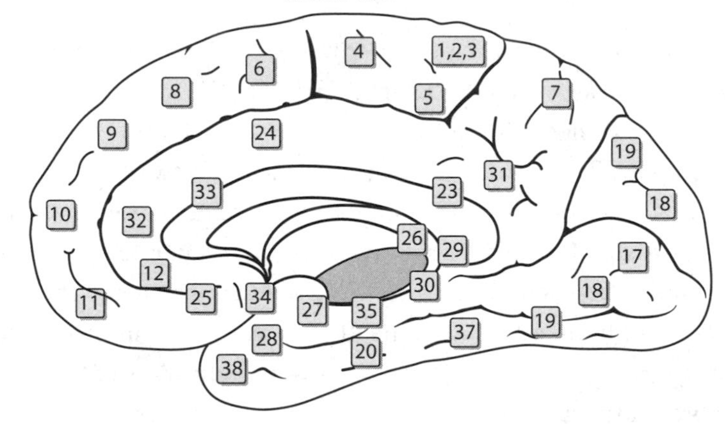

FIGURE 0.2 Brodmann brain areas.

the brain areas, evolved. Four white matter tracts that have evolved most recently connect the brain areas that have evolved most recently, the same brain areas that are associated with the cognitive skills discussed in this book. As will be detailed in subsequent chapters, what is known about brain evolution and what is known about the acquisition of specific cognitive skills fit together remarkably well.

The importance of the brain's connecting fibers in making us uniquely human also suggests that there is no single "god part" of the brain. Like almost all human higher cognitive functions, thoughts about gods are the product of a *network* of multiple brain areas. Such networks have been described as "grids of connectivity" that "allow a very large number of computational options to be associated with specific cognitive processes." These networks have also been referred to as "modules" or "cognitive domains." Thus, even language, which has traditionally been thought to be localized in two brain areas (Broca's and Wernicke's areas), is now known to be part of a network involving at least five other areas. Therefore, there is no "god part of the brain," but there is a network that controls thoughts about gods and religious beliefs. This is the network of the numinous, the same network that controls the cognitive skills that make us uniquely human.[10]

THE NATURE OF THE EVIDENCE

Since the proposed evolutionary theory of this book depends on an understanding of how the brain evolved, it is reasonable to ask how we know what we know. What is the nature of the evidence? Information regarding hominin brain evolution comes from five major research areas: studies of hominin skulls; studies of ancient artifacts; studies of postmortem brains from humans and primates; studies of brain imaging of living humans and primates; and studies of child development.

Hominin skulls have been an important source of information on human brain evolution. It would of course be preferable to have the brains themselves, but following death, the brain is one of the first organs to deteriorate, liquefying within hours if the temperature is warm. We therefore have no brains from ancient hominins to examine. Imagine how much we could learn if we had preserved brains from *Homo habilis*, *Homo erectus*, *Homo neanderthalensis*, and early *Homo sapiens* to place side by side, to compare with the brain of modern *Homo sapiens* and then to dissect each brain in minute detail.

Alas, we do not. What we do have, however, are skulls that held those brains. Like Hamlet standing in the churchyard with the skull of "poor

Yorick," we can use the skulls to speculate on past behaviors that were the products of the brains within. As the developing brain grows during fetal and infant life, the pliable bones of the skull mold themselves to the shape of the brain. Skulls are thus like ancient footprints left in volcanic ash that subsequently hardened; we no longer have the feet to examine, but we do have the shape of the feet and even some details of the toes.

Skulls that have been well preserved can provide us with considerable data. The volume of the brain is, of course, relatively easy to calculate. The overall shape of the brain is also evident, including whether the two halves are symmetric, as they were early in hominin evolution but not later. By examining the shape, we can also make informed guesses regarding the relative size, and thus importance, of the frontal, parietal, temporal, and occipital areas. In early hominin brains, the occipital area was prominent, but in later brains other areas became more developed. The inner lining of skulls includes grooves for the major arteries and veins, and on the floor of the skull are concave impressions for the cerebellum and the underside of the frontal lobes. On especially well-preserved skulls, it is even possible to find impressions of individual brain ridges, or gyri. Overall, having the skull is a distant second choice to having the brain to examine, but when combined with other evidence of our ancestors' behavior, the skull nonetheless may yield substantial useful information.

Ancient artifacts are a second important source of clues regarding the cognitive abilities and behavior of earlier hominins, and thus the evolution of the brain. The finding of improved tools made by *Homo habilis* two million years ago suggests higher intelligence and improved cognitive function in general. As previously noted, the finding of shells fashioned for self- ornamentation and used by early *Homo sapiens* approximately 100,000 years ago suggests that they had acquired an ability to think about what others were thinking about them. The finding of food, tools, weapons, jewelry, and other supplies buried with dead bodies by modern *Homo sapiens* approximately 27,000 years ago suggests that they had acquired an ability to think about a possible life after death.

THE DATING OF SKULLS AND ARTIFACTS

Ancient skulls and artifacts are useful for understanding human evolution, however, only insofar as they can be dated with reasonable accuracy. For the period up to about 40,000 years ago, radiocarbon dating has been commonly used. Carbon is present in all living things, and an isotope, carbon-14, decays at a predictable rate. By measuring the amount of carbon-14 remaining in a sample of hair, bone, wood, charcoal, or other organic matter, it is possible to calculate a probable date with a margin of error of approximately 10 percent. Thus, a burial that is radiocarbon dated to 30,000 years ago probably took place between 27,000 and 33,000 years ago. A limitation of radiocarbon dating is that the amount of carbon-14 in the atmosphere has varied over time, depending on solar activity and the earth's magnetic field, so various methods have been developed to correct for this source of error. Because of such limitations, radioactive thorium and uranium are now increasingly being used as an alternative dating method.

For years earlier than 40,000 years ago, dating is much less precise. Various methods have been used, including a system measuring the decay of potassium to radioactive argon (potassium-argon dating), a system measuring the buildup of electrons due to radioactive damage (electron spin resonance dating), and a system based on DNA mutations. All three systems have very wide margins of error, and the earlier the event being dated, the wider the margin of error. DNA mutations, for example, have been used to estimate when species split, such as the ancestors of chimpanzees from the earliest hominins. It was recently discovered that DNA mutations occur more slowly than previously assumed. Thus, the chimpanzee-hominin split, which was thought to have occurred four to seven million years ago, may actually have occurred eight to 10 million years ago. And the migration of early *Homo sapiens*

(*continued*)

out of Africa, commonly dated to about 60,000 years ago, may instead have occurred 120,000 years ago. Thus, all dates discussed in this book before 40,000 years ago should be assumed to have wide margins of error.

J. Hellstrom, "Absolute Dating of Cave Art," *Science* 336 (2012): 1387–1388; A. Gibbon, "Turning Back the Clock: Slowing the Pace of Prehistory," *Science* 338 (2012): 189–191.

A third major research resource for learning about brain evolution is postmortem brains from humans and primates. It is generally accepted that brain areas that evolved early during the evolution of *Homo sapiens* also mature early in the development of an individual; similarly, areas that evolved later mature later. As summarized in one study of this phenomenon, "phylogenetically older cortical areas mature earlier than the newer cortical regions." For example, brain areas associated with specific muscle functions, such as the movement of arms, lips, and tongue, were among the earliest areas to evolve and are also among the earliest areas to mature, thus enabling a newborn to grasp and suckle from its mother's breast.[11] Three methods of assessing the relative maturation of brain areas are summarized in appendix A.

In addition to providing information on brain areas that developed more recently in evolution, postmortem brains of humans can also be compared with the postmortem brains of chimpanzees and other primates. Such comparison studies reveal which hominin brain areas have increased or decreased in size over the course of evolution, the relative degree of connectivity of various brain areas, whether there are unusual cell types specific to hominins, the anatomical spacing of the cells, and whether there are differences in the chemical composition of such things as neurotransmitters and proteins.

Regarding the size of specific areas, it is generally assumed in brain development that the size of a specific brain area correlates with the importance of the function served by that area. This principle has been summarized as follows: "The mass of neural tissue controlling a particular function is appropriate to the amount of information processing involved in performing the function." Thus, a bat, which relies on sound, has a large auditory cortex; a monkey, which relies on vision, has a large visual cortex; a rat, which relies on smell, has a large olfactory cortex; and a desert mouse, which relies on memory to recall where it has hidden seeds, has a highly developed memory area (hippocampus). Studying the relative size of specific human brain areas in comparison with those of chimpanzees can therefore help to identify which human areas are most important and have evolved more recently.[12]

In addition to studying hominin skulls, artifacts, and human postmortem brains, a fourth approach to understanding how brains evolved is to study living brains using recently developed imaging techniques. Such techniques include magnetic resonance imaging (MRI) and its functional component (fMRI) as well as diffusion tensor imaging (DTI), which is especially useful for assessing connections between brain areas. MRI studies of living humans and chimpanzees have highlighted structural brain differences between them and thus complemented the postmortem studies. MRI studies of children have also been used to assess which brain areas mature early and which ones later. The results have been remarkably consistent and show that "phylogenetically older brain areas mature earlier than newer ones." The combination of the postmortem studies and the MRI studies indicates which brain areas developed most recently during the course of hominin brain evolution.[13]

Functional MRI (fMRI) studies can also be used to link specific brain functions to specific brain areas or networks. For example, an individual can be asked to think about what another person is thinking while an fMRI measures which brain areas are activated. This then links the process of thinking about another person to the activity of specific brain areas. Since we know which brain areas developed more recently in evolving hominin brains, the fMRI studies give us information regarding the functions of the more recently developed brain areas.

The use of diffusion tensor imaging (DTI), which has recently become available, has enabled us for the first time to visualize the brain's white matter connecting tracts in living individuals. To date, over 15 separate connecting tracts have been identified, and by doing DTI studies on children and young adults, it is possible to assess the degree of maturation of each tract at different ages. Some connecting tracts are mature shortly after birth; an example is the corpus callosum, the large tract connecting the two hemispheres. This tract has also been linked to intelligence and found to have been especially large in a postmortem study of the brain of Albert Einstein. Another white matter tract that matures shortly after birth is the inferior longitudinal fasciculus, which connects the prefrontal brain area with the occipital lobe and visual cortex at the back of the brain. By contrast, four other white matter connecting tracts are among the very last to mature, and these connect the parts of the brain that are crucial for becoming modern *Homo sapiens*. These four connecting tracts are the superior longitudinal fasciculus, arcuate fasciculus, uncinate fasciculus, and cingulum; they are shown in figure 0.3 and will be discussed in subsequent chapters.[14]

The fifth and final major resource that is useful for studying the evolution of the human brain is the cognitive development of children. For many years, it was thought that the physical development of human fetuses precisely mirrored the evolutionary development of the species. Thus, human fetuses were said to have a tail and pharyngeal pouches that resembled the tail and gill slits of ancient vertebrates from which mammals evolved. Based on such observations, generations of biology students were taught that "ontogeny [the physical development] recapitulates phylogeny [the evolutionary development]."

The rigid interpretation of this maxim has been discredited by Harvard University biologist Stephen Jay Gould and others. The *physical* development of individuals does not precisely recapitulate the evolutionary development of the species. However, there are broad parallels, and this appears to be especially true for human *cognitive* development. Sir John Eccles, a British neuroscientist and Nobel laureate who devoted his life to the study of mammalian brains, believed that "the progressive development from the consciousness of the baby to the self-consciousness in the child provides a good model for the emergent evolution of self-consciousness in the hominids." Child development specialist Jean Piaget also believed that "the development of thought in children closely

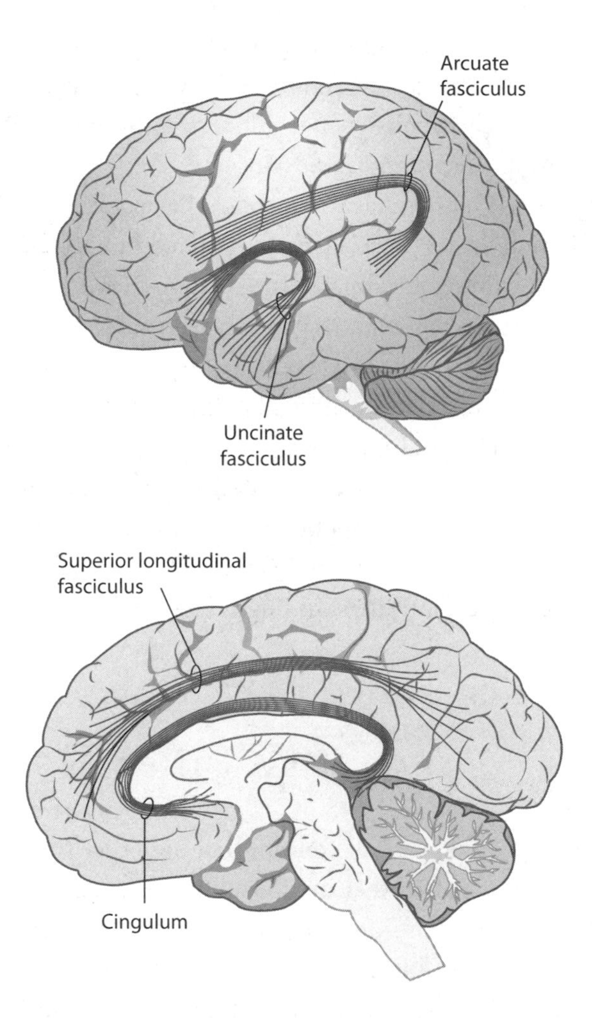

FIGURE 0.3 White matter connecting tracts important for skills making us uniquely human.

parallels the evolution of consciousness in our species." More recently, Daniel Povinelli, a University of Southwestern Louisiana psychologist who has specialized in comparing chimpanzee and human cognitive processes, noted that "comparing the ontogeny of [human] psychological capacities should allow evolutionary psychologists to reconstruct the order in which particular features of mental state attribution evolved." A symposium on this subject concluded that "the sequence of cognitive development in humans roughly parallels the sequence of its evolution in ancestral forms." Thus, the cognitive development of children can be used as a clue to help reconstruct evolutionarily the cognitive development of hominins, including *Homo sapiens*.[15]

Although much has been learned about the evolution of the human brain, much more is yet to be learned. As stated in a recent assessment of the field, "our understanding of the relation between the structure and function of the brain remains primitive, especially when compared to other organ systems." The broad outlines are reasonably clear, but the finer details of brain evolution are still being sorted out. Over the next decade, we can expect additional progress as brain neuroimaging techniques become increasingly sophisticated. We will then have a much better understanding of the function of specific brain networks as well as the evolution of the connecting white matter tracts, which should lead us to an even better understanding of the emergence of the gods.[16]

PARALLEL EVOLUTION

There is one additional critical concept that underlies the argument of this book. When it is said that neurons, glial cells, and brain connections evolved over millions of years, what is actually meant? Genes are stretches of DNA and can be altered by a number of factors, including errors in cell division, radiation, viruses, and some chemicals. Evolution of a brain occurs when the molecular structure of a gene associated with the brain undergoes an alteration that provides the organism with some reproductive advantage. For example, when *Homo sapiens* acquired what is called "autobiographical memory," as will be described in chapter 5, they were able to plan for the future much more skillfully than other hominins living at that time. Some altered genes are disadvantageous to the organism, and these genes die out. Other altered genes provide some reproductive advantage, and these genes are more likely to be passed on. Evolution is thus figuratively the attempt of genes to get ahead in life. Darwin called this process natural selection:

> It may metaphorically be said that natural selection is daily and hourly scrutinizing, throughout the world, the slightest variations; rejecting those that are bad, preserving and adding up all that are good; silently and insensibly working, whenever and wherever opportunity offers, at the improvement of each organic being in relation to its organic and inorganic conditions of life. We see nothing of these slow changes in progress, until the hand of time has marked the lapse of ages.

Our brains are thus the remodeled products of 200 million years of such trial-and-error natural experiments. It should therefore not surprise us to find that our brains include many features of unintelligent design, features that make no sense today but that probably evolved to prevent some ancestral mammals from becoming an hors d'oeuvre for a brontosaurus.[17]

A special aspect of evolution that is important for the theory being proposed in this book is the existence of parallel evolution. This occurs when organisms that have had a common genetic origin continue to evolve along similar lines even though they have been separated from one another for thousands, or even millions, of years. The separated organisms develop along similar lines either because they are subjected to similar external selection pressures, such as climate or food supply, or because they have internal constraints, such as common anatomical structures that limit the number of developmental possibilities. Parallel evolution has been defined as "the recurrent tendency of biological organization to arrive at the same 'solution.'" The products of parallel evolution have both intrigued and perplexed observers. Harvard historian Daniel Smail called them "some of the eeriest features of Postlithic human society. . . . Agriculture was independently invented on different continents, as were writing, pottery, priestly castes, embalming, astronomy, earrings, coinage, and holy virginity. . . . We celebrate the diversity of human civilizations, but it is the similarities that are the most startling." Such phenomena become comprehensible if understood as products of continuing brain evolution.[18]

EXAMPLES OF PARALLEL EVOLUTION

The most widely cited example of parallel evolution is the evolution of mammals in Australia. More than 100 million years ago, the continents drifted apart and Australia became isolated from other continents. However, Australian mammals and the mammals on other continents had shared common ancestors before the continents drifted apart, so some of the descendants continued to evolve along remarkably similar lines. Examples of such parallel evolution include the Australian crest-tailed marsupial mouse and the European mole, the Australian sugar glider and the North American flying squirrel, and the Tasmanian wolf and the North American gray wolf. There are, of course, other mammals that evolved along different lines because of genetic mutations and different external selection pressures, such as climate, food supply, predators, or other factors.

Studies that have compared the brains of marsupial mammals in Australia with the brains of placental mammals on other continents have demonstrated the anatomical underpinnings of parallel evolution. The brain areas governing vision, hearing, and sensory stimuli are said to be remarkably similar in both types of mammals. The authors of one study concluded that "marsupials have evolved an array of morphological, behavioral, and cortical specialization that are strikingly similar to those observed in placental mammals occupying similar habitats, which indicate that there are constraints imposed on evolving nervous systems that result in recurrent solutions to similar environmental challenges."

Another example of parallel evolution in brain development comes from a study comparing the brains of Old World and New World monkeys, which have evolved separately for 30 million years. One New World monkey, the cebus, uses a precision grip in which "the thumb and forefinger are brought into contact with one another to manipulate small objects or engage in goal-directed tool use." An Old World monkey, the macaque, also uses a precision grip. When the brains of both monkeys were examined, remarkable anatomical similarities were found in the part of the parietal lobe that governs hand use. The authors concluded that "evolutionary changes involving skeletal, muscular and neural features proceed in parallel and, thus, features of the body and brain are linked. . . . The similarity of these [anatomical] fields in cebus monkeys and distantly related macaque monkeys with similar manual abilities indicates that the range of cortical organizations that can emerge in primates is constrained, and those that emerge are the result of highly conserved developmental mechanisms that shape the boundaries and topographic organizations of cortical areas."

J. Karlen and L. Krubitzer, "The Functional and Anatomical Organization of Marsupial Neocortex: Evidence for Parallel Evolution Across Mammals," *Progress in Neurobiology* 82 (2007): 122–141; J. Padberg, J. G. Franca, D. F. Cooke et al., "Parallel Evolution of Cortical Areas Involved in Skilled Hand Use," *Journal of Neuroscience* 27 (2007): 10106–10115.

Parallel evolution of brain development can explain many of the remarkably similar developmental trajectories described in this book. For example, it seems likely that the initial genetic brain changes that enabled us to fully place ourselves into the past and future (autobiographical memory) took place before *Homo sapiens* left Africa. Because these brain developments were already underway, *Homo sapiens* continued to evolve cognitively along roughly similar lines for thousands of years whether they ended up in Portugal, Pakistan, Peru, or Papua New Guinea. Insofar as the widely disparate groups experienced similar selection pressures, such as increasing population pressures following the domestication of plants and animals, it should not surprise us to find disparate geographical groups arriving at similar outcomes. For example:

- About 40,000 years ago, the first examples of visual arts appeared in cave paintings in places we now call Spain and Indonesia, and in sculpted ivory figurines in Germany.

- Between 11,000 and 7,000 years ago, plants and animals were independently domesticated in southwest Asia, China, Papua New Guinea, Peru, and probably Mesoamerica.
- By about 9,000 years ago, ancestor worship had apparently become widespread in both southwest Asia and China.
- Between 6,500 and 5,000 years ago, higher gods had independently emerged in southwest Asia, China, and probably Peru.

Psychologists Mark Leary and Nicole Buttermore similarly suggested that "the neurological substrates for conceptual-self ability were in place before *H. sapiens* began to disperse out of Africa. . . . This may reflect a case of parallel evolution in which cognitive changes that occurred before the dispersion from Africa had evolutionary momentum."[19]

Thus, the cognitive evolution of the human brain made possible the emergence of gods and civilizations. This would then be the starting point for a remarkable period of human development. In a mere 6,000 years, we would go, in the words of brain researcher Marcel Mesulam, "from the oxcart to Voyager, from the Sphynx to Rodin's Kiss, and from Gilgamesh (by way of the Odyssey) to the Divine Comedy." It has been a truly extraordinary journey. But to fully understand how all of this came about, we must begin at the beginning, with the first of the five major cognitive advances.[20]

1

THE MAKING OF THE GODS

1

HOMO HABILIS

A Smarter Self

The history of religious belief is rarely given centre stage in grand narratives of the evolution of civilization and of humanity and yet the urge to comprehend the human condition—the quest for soul food—may be just as great as the quest for food and reproductive success.

—MIKE PARKER PEARSON, *THE ARCHEOLOGY OF DEATH AND BURIAL*, 1999

The gods were born following a pregnancy lasting approximately two million years. It took that long for hominin brains to evolve structurally and functionally from being primate-like brains to being brains that possessed the cognitive faculties of modern *Homo sapiens.* Insofar as an evolutionary origin of deities is correct, the concept of a god would not have occurred to hominins prior to about 40,000 years ago, and the gods themselves would probably not have become fully visible prior to about 10,000 years ago. The human brain, and thus the self-aware human world, would not have been ready for them before that time.

Mammalian brains had, of course, been evolving for 200 million years prior to that time. For the first 140 million years of their existence, mammals were insignificant "small creatures living in the nooks and crannies of a dinosaur's world." During those eons, evolution was experimenting with the development of the three-part brain—the forebrain, midbrain, and hindbrain—that forms the central nervous system chassis for all mammals.[1]

About 65 million years ago, an asteroid apparently struck earth, producing a cataclysm that killed the dinosaurs and many other creatures. Mammals not only survived but thrived in a world now devoid of Jurassic predators. As Stephen Jay Gould noted, "We must assume that consciousness would not have evolved on our planet if a cosmic catastrophe had not claimed the dinosaurs as victims. In an entirely literal sense, we owe our existence, as large and reasoning mammals, to our lucky stars." Our origin, added Gould, makes *Homo sapiens* "a kind of cosmic accident, just one bauble on the Christmas tree of evolution."[2]

With the disappearance of dinosaurs, mammals rapidly diversified, grew larger, and became the new lords of the earth. The mammalian forebrain increased disproportionately in size compared to the midbrain and hindbrain and eventually occupied most of the space within the skull. As the forebrain grew, it differentiated into the four lobes (frontal, temporal, parietal, and occipital), basal ganglia, hippocampus, amygdala, thalamus, and hypothalamus. Most significantly, the brain developed a thin layer called the neocortex, which has been said to be like a 13-inch pizza covering the four lobes of the brain. According to Georg Striedter's *Principles of Brain Evolution*, "The neocortex was the key innovation of mammalian brains," because it included six layers of neurons, compared to the three layers in the cortex of earlier animals. Since neurons are connected three-dimensionally, both horizontally and vertically, to other neurons, the additional three layers increased neuronal connections exponentially, thereby making possible the processing of much more complex information and thought.[3]

As part of the diversification of mammals, the first primates appeared approximately 60 million years ago. They proliferated rapidly into hundreds of species, of which 235 species still exist. About 30 million years ago, a group known as New World monkeys (for example, cebus monkeys and marmosets) went their separate evolutionary way, and 25 million years ago the Old World monkeys (for examples, baboons and macaques) did the same thing. The great apes, the group most closely related to us, began dividing about 18 million years ago, with the orangutan, and then the gorilla, starting down separate evolutionary paths. Finally, about six million years ago, the hominins separated from chimpanzees, our closest hominid ancestor.

It is important to note that the hominins did not evolve from chimpanzees as we know them. Rather, both hominins and chimpanzees evolved from a common ancestor that lived about six million years ago. During the intervening time, both the hominin line and the chimpanzee line continued to evolve. Among the chimpanzees, for example, one group became geographically isolated in West Africa about two million years ago, and that group evolved into bonobos, also called pygmy chimpanzees. Insofar as the evolving chimpanzee line was subjected to similar evolutionary pressures as the evolving hominin line was during the 6 million years, it would not be surprising, given the principles of parallel evolution, to find that chimpanzees would develop some cognitive abilities similar to those developed by hominins. Awareness of self, to be discussed in chapter 2, is an example of such parallel development.

THE FIRST HOMININS

The evolution of one species into separate species is usually a gradual process. Thus, *Sahelanthropus tchadensis*, a fossil found in 2001 in Chad and thought to be at least six million years old, has been classified by some as the first bipedal hominin but by others as a chimpanzee. Its brain capacity was less than 400 cubic centimeters, equal in size to the brain capacity of modern chimpanzees.[4]

Sahelanthropus tchadensis was followed during the next four million years by *Ardipithecus kadabba*, *Ardipithecus ramidus*, and several species classified as *Australopithecus—anamensis*, *afarensis*, *africanus*, *garhi*, *boisei*, *robustus*, *aethiopicus*, and, from fossils discovered in 2010, *sediba*.

There is much discussion regarding which hominin descended from which other hominin, but in fact there are not yet a sufficient number of specimens to make such determinations with any certainty. The study of early hominin fossils has been said to still be in "the stamp-collecting phase that begins most branches of science."[5]

What is clear is that these early hominins had a brain capacity of approximately 400 to 475 cubic centimeters, only slightly larger than that of chimpanzees, and their behavior was quite similar to that of chimpanzees. They spent their days foraging for fruits, nuts, roots, and tubers and retreated to trees to escape predators and to sleep. Some researchers have claimed that some species of Australopithecus used stone tools, but other researchers have been doubtful. The most famous examples of Australopithecus are "Lucy," whose fossils were found in 1974 in Ethiopia, and three sets of footprints embedded in volcanic ash in Tanzania. We occasionally romanticize Australopithecus and tell ourselves that they were not very different from us, but except for walking upright, they were in fact very different. Because of their rudimentary brain development, they could not think about themselves, they could not boast of their accomplishments, they could not gossip about other australopithecines, they did not worry about what might happen after they died, and they did not worship gods. Thus, it is generally believed that Australopithecus individuals "differed from other African apes (just as the other African apes differed from one another) but they were still apes, in mind if not in body."[6]

When *Homo habilis* evolved approximately two million years ago, the world of early hominins became significantly more interesting, because of both its brain size and its behavior. *Homo habilis* is generally regarded as being the first hominin to have diverged significantly from its primate ancestors, although its precise relationship to other early members of the Homo species—such as *Homo rudolfensis*, *Homo ergaster*, and the recently discovered *Homo naldi*—is far from settled. Fossils of *Homo habilis* have been discovered in Ethiopia, in northern Kenya, and especially in the Olduvai Gorge in Tanzania, which Louis and Mary Leakey made famous.

Homo habilis is thought to have lived between 2.3 and 1.4 million years ago, although recent finds in Ethiopia suggest that it may have existed as early as 2.8 million years ago. Its average brain size is estimated to have been about 630 cubic centimeters and thus to have been one-third larger than the brain of Australopithecus.

The larger brain of *Homo habilis* made it smarter than Australopithecus, and it demonstrated this intelligence by making crude stone tools. This was mostly done by breaking rocks to produce sharp stone edges. Crude stone tools have been found dated to 3.3 million years ago, but those made by *Homo habilis* were more sophisticated. These have been found in abundance in association with *Homo habilis* fossils. Although crude, such tools would have been effective for cutting the hides and tendons of dead animals, thus allowing the tool-user to strip meat. The stone tools could also have been used to break open animals' long bones and extract the marrow, an especially rich source of protein. Animal bones found in association with the stone tools suggest that the tools were used in this way. The bones also suggest that *Homo habilis* was probably a meat eater, in contrast to earlier hominin species. There is no evidence that *Homo habilis* hunted animals, so they probably scavenged for animals that had been killed by other animals or had died of old age or disease.

The use of tools is, of course, not unique to hominins. Many birds have been observed using tools, including crows, which use sticks and carefully cut leaves to extract insects from holes, and Egyptian vultures, which drop stones on ostrich eggs to crack them open. Sea otters use stones to break the shell of snails and crabs. Monkeys have been observed using sticks to kill snakes and rocks to crack open oyster shells, and it is well known that chimpanzees use sticks, from which they strip the leaves, to forage in termite mounds, and stones to crack open nuts.

What makes the stone tools used by *Homo habilis* different is their complexity. According to Cambridge University archeologist Steven Mithen: "To detach the type of flakes one finds in the sites of Olduvai Gorge, one needs to recognize acute angles on the [stone] nodules, to select so-called striking platforms and to employ good hand-eye coordination to strike the nodule in the correct place, in the right direction and with the appropriate amount of force."[7]

Attempts have been made to teach chimpanzees and bonobos to make stone tools similar to those made by *Homo habilis*. One especially clever bonobo, rewarded by food treats, successfully made stone tools, but they were significantly inferior to those of *Homo habilis*. According to Mithen, the bonobo "never developed the concept of searching for acute angles . . . or controlling the amount of force in percussion." Mithen speculated that *Homo habilis* had already developed cognitive skills superior to those of modern chimpanzees, "an intuitive physics in the mind . . . perhaps even a technical intelligence." Such cognitive superiority is supported by evidence that *Homo habilis* occasionally used one tool to make another tool, such as using a stone flake to sharpen a stick; this behavior is unknown among chimpanzees.[8]

Additional evidence of the intelligence of *Homo habilis* includes the fact that they traveled several miles to obtain specific types of stones superior for use as tools. They also carried stone tools to new sites, evidence of planning and anticipation of future use. Archeologist Kenneth Feder of Central Connecticut State University said that such behavior suggests "a high level of planning and intelligence." Such planning and storage of tools for future use are occasionally found among chimpanzees. An adult male chimpanzee in a Swedish zoo, for example, regularly collected and stored stones prior to the zoo's opening time, which he then used to throw at spectators across the moat surrounding his enclosure.[9]

Thus, what was *Homo habilis* really like? They possessed advanced physical skills and some ability to plan and were clearly smarter than their hominin ancestors. However, despite their greater intelligence, there is no evidence that they possessed self-awareness or any of the other higher cognitive functions that would distinguish later hominins and lead to the emergence of the gods. British psychologist Nicholas Humphrey painted a hypothetical picture of what *Homo habilis* was like:

> Once upon a time there were animals ancestral to man who were not conscious. That is not to say that these animals lacked brains. They

> were no doubt percipient, intelligent, complexly motivated creatures, whose internal control mechanisms were in many respects the equals of our own. But it is to say that they had no way of looking in upon the mechanism. They had clever brains, but blank minds. Their brains would receive and process information from their sense-organs without their minds being conscious of any accompanying sensation, their brains would be moved by, say, hunger or fear without their minds being conscious of any accompanying emotion, their brains would undertake voluntary actions without their minds being conscious of any accompanying volition. . . . And so these ancestral animals went about their lives, deeply ignorant of an inner explanation for their own behaviour.[10]

"Clever brains but blank minds" appears to capture the essence of *Homo habilis.*

THE BRAIN OF *HOMO HABILIS*

Why was *Homo habilis* smarter than its predecessors? One reason, quite simply, is that its brain was more than 50 percent larger than the brains of its predecessors. Although four million years had passed since the earliest hominins and chimpanzees had separated from their common ancestor, during that time hominin brains had grown only slightly larger than chimpanzee brains. Suddenly, two million years ago, hominin brains began growing much more rapidly, initiating a growth pattern that would eventually lead to the oversized brains of *Homo sapiens*, brains that have been characterized as "freakishly large for a mammal of our body size." Specifically, "the human brain is 3.5 times bigger than expected for an ape our size." Philip Tobias, a South African paleoanthropologist who did much of the original research on *Homo habilis* skulls and named the species, noted that "it was with *H. habilis* that there began the remarkably disproportionate enlargement of the brain that is one of the hallmarks of mankind." Similarly, Michael Rose, an evolutionary biologist at the University of California, claimed that "the expansion of the human brain over the last 2 million years is one of the most rapid and sustained such morphological developments known in the fossil record."[11]

As a general rule, when it comes to brains, bigger is better. Tobias, for example, estimated that the increased brain size of *Homo habilis* resulted in an additional one billion neurons compared to the brain of Australopithecus. But size is not everything, since it is known that the brains of highly intelligent and accomplished people may vary considerably in size. For example, the brains of English satirist Jonathan Swift and Russian novelist Ivan Turgenev each weighed over 2,000 grams, whereas the brain of French novelist Anatole France weighed just 1,000 grams. As will be described in chapter 3, the brains of Neandertals were as big, and often bigger, than the brains of modern *Homo sapiens*. And, as noted in Appendix A, elephants have brains four times larger, and whales five times larger, than humans. However, if brain size is scaled to body size, human brains are among the largest known. Chimpanzees, for example, weigh about the same as humans, but chimpanzee brains are less than one-third the size of human brains. This contrasts with other organs, such as the heart, lungs, liver, and kidneys, which are of similar sizes in chimpanzees and humans. A large brain is thus a distinguishing feature that sets humans apart from other primates, but brain size alone is not what makes humans unique.[12]

The uniqueness of the human brain, rather, lies in the specific areas of the brain that are enlarged and in the intensity of the connections among those areas. According to Tobias, the skulls of *Homo habilis* suggest that "the increase in cerebral substance . . . is evident mainly in the frontal and the parietal lobes" and "appears to be less marked" in the temporal and occipital lobes. Specifically, in the frontal lobe there appears to be "quite marked remodeling of the lateral parts of the frontal lobe," and in the parietal lobe the superior parietal lobule and inferior parietal lobule are both "especially well developed." Tobias concluded that with the brains of *Homo habilis*, "hominid evolution attained a new level of organization."[13]

Thus, two facts seem well established. First, *Homo habilis* appears to have been smarter than its predecessors, and second, its brain had grown larger disproportionately in the frontal and parietal regions. It is reasonable to assume that these two facts may be causally connected, but is there any data to support this?

In fact, there is. In recent years there has been an outpouring of neuroimaging studies attempting to localize the components of intelligence in the human brain. A summary of 37 such studies reported a "striking consensus" in localizing intelligence to a network involving areas in the frontal and parietal regions and the connections between these regions. Thus, the results of neuroimaging studies localizing intelligence in the brains of contemporary *Homo sapiens* coincide nicely with the brain areas that grew disproportionately large in *Homo habilis* two million years ago at the same time that *Homo habilis* was becoming more intelligent.[14]

BASIC AREAS ASSOCIATED WITH INTELLIGENCE

The specific brain areas that become activated in neuroimaging studies of intelligence differ slightly depending on the test that is utilized, as would be expected. For example, many studies have used the Wechsler Adult Intelligence Scale (WAIS), a test that measures verbal comprehension, perceptual organization, processing speed, and working memory, defined as short-term memory needed to solve immediate problems. The brain areas activated by this intelligence test include the following frontal areas: frontal pole (BA 10), lateral prefrontal cortex (BA 9 and 46), and anterior cingulate (BA 24 and 32). The inferior parietal lobe (BA 39 and 40) is also activated by the WAIS. When other tests of intelligence are used, such as the measurement of blood brain flow while people are playing chess, another frontal area (premotor cortex, BA 6) and another parietal area (superior parietal, BA 7) are also prominent (figure 1.1). The authors of these studies concluded that "there is much

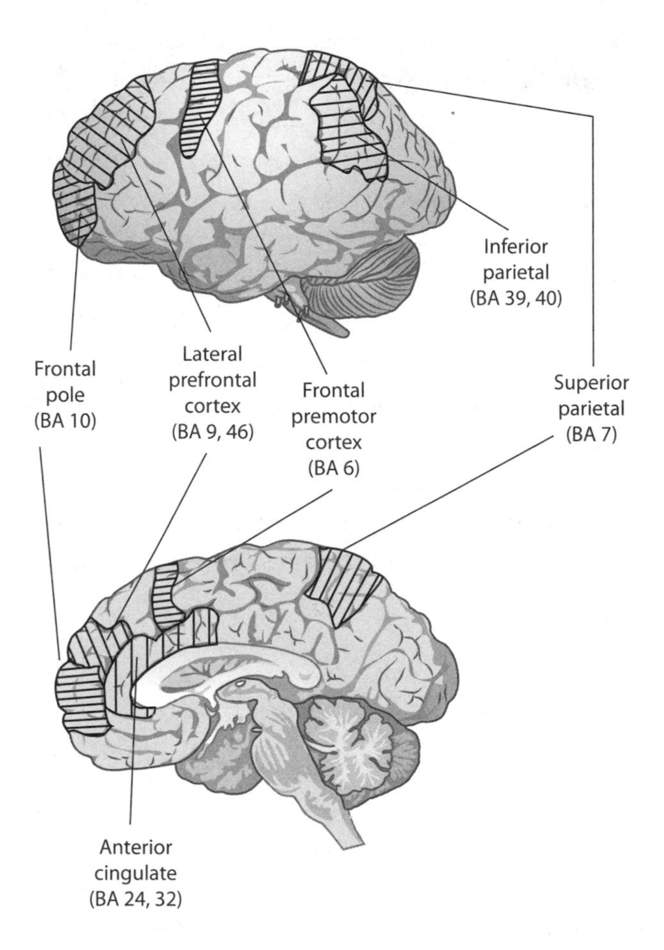

FIGURE 1.1 *Homo habilis*: a smarter self.

neuroanatomical consistency among results, which we have described as defining a specific frontal-posterior [parietal] network."[15]

What is known about these brain areas that are apparently associated with intelligence?

First, they are almost all part of Flechsig's "terminal zones" and thus evolutionarily are thought to have developed more recently. In fact, the frontal pole (BA 10) and lateral prefrontal cortex (BA 9 and 46) were classified by Flechsig as being the last brain areas to have evolved. Second, most of the brain areas that have expanded the most are what are known as association areas; such areas are involved not in more simple muscle or sensory function but rather in complex brain functions, such as assessing input from multiple other brain areas and coordinating appropriate responses. For example, if *Homo habilis* had put his hand behind a rock and simultaneously heard a hissing noise and felt a slimy creature, his association areas would have integrated these sensory inputs and directed an instantaneous withdrawal of his hand. It is the association areas of the brain, not the primary brain areas, that have evolved most recently and produced the unique cognitive skills of *Homo sapiens*. This principle was clearly illustrated by Todd Preuss, a neuroscientist and primatologist at Emory University who has extensively compared primate and human brains. He concluded that "the primary areas maintained approximately ape-like sizes in human evolution, while association cortex underwent enormous expansion." For example, when Preuss compared the size of the human primary motor or visual cortex with the analogous areas in primate brains, the human brain areas were not larger than expected. By contrast, when Preuss compared the size of human association areas with the analogous areas in primate brains, the human areas were many times larger than expected.[16]

Other studies also support specific parts of the frontal and parietal lobes as being critical for intelligence. The frontal pole (BA 10), for example, is said to have "more expanded than any other part in the human brain as compared to our ancestors." It plays a critical role in information processing, working memory, social cognition, the processing of emotions, and the planning of future actions. A recent study that compared the spacing of neurons in the frontal pole of humans with the frontal pole of great apes reported that the spacing of neurons in the former afforded greater neuron interconnectivity. The relative importance of the human frontal pole can also be assessed by the fact that it contains more than four times more neurons than does the similar brain area in chimpanzees. The frontal premotor cortex (BA 6) has many functions, including being activated by tasks involving the abstracting of rules and associative learning. The lateral prefrontal cortex (BA 9 and 46) plays a central role in executive functions, including planning and reasoning, as will be discussed in chapters 4 and 6.[17]

The superior parietal area (BA 7), also known as the precuneus, performs a wide variety of cognitive, sensory, and visual functions. Both the superior and the inferior (BA 39 and 40) parietal areas play important roles in intelligence and have also been linked to other intellectual functions, such as deductive reasoning. And it is probably not a coincidence that when Albert Einstein's brain was examined after his death, his inferior parietal area was found to be "15 percent wider than controls." This is an area that integrates visual imagery with mathematical thinking and other cognitive skills. In the previous chapter, it was noted that Einstein's corpus callosum, which connects the two hemispheres of the brain, was also enlarged; the part of the corpus callosum that was most enlarged was the part that connects the inferior parietal areas in both hemispheres. Thus, an enlarged inferior parietal area and its connecting fibers may have been one reason for Einstein's intellectual prowess.[18]

As the frontal and parietal brain areas were increasing in size in *Homo habilis*, it is likely that the white matter tracts connecting these two parts of the brain were also developing. The major connections are three tracts that together make up the superior longitudinal fasciculus. The three tracts connect the prefrontal cortex to the superior parietal (BA 7), inferior parietal angular gyrus (BA 39), and inferior parietal supramarginal gyrus (BA 40), respectively. Studies of the maturation of the superior longitudinal fasciculus have reported it to be "one of the slowest maturing white matter tracts," thus consistent with the development of greater intelligence over the past 4 million years. According to other studies, the superior longitudinal fasciculus "can be clearly identified only in highly developed species. . . . This strongly suggests involvement of SLF [superior longitudinal fasciculus] in high-level brain functions." As noted previously, it is not just the existence of white matter connecting fibers that is important but also the speed at which they conduct information. This is especially important for intelligence. For example, a study of comparative intelligence among primates and other animals reported that the two most important predictors of intelligence were the number of neurons in the brain and conduction velocity of the connecting tracts.[19]

WHY DID THE BRAIN INCREASE IN SIZE?

The rapid increase in hominin brain size beginning approximately two million years ago raises two questions: How did the increase occur, and why did the brain begin to increase in size at that time, after having remained relatively constant in size for the preceding four million years. Regarding the first question, there has been an ongoing debate among scientists about whether hominin brains grew larger simply by renovating existing brain areas or whether they grew larger by creating new brain areas. By way of analogy, did the house down the street become larger because the owners enlarged the existing rooms or because they built new rooms as additions?

Although the question is still unresolved, there is a consensus that most brain evolution occurs by the former method; in other words, "opportunistic evolution has conscripted old parts of the brain to new functions in a rather untidy fashion." It seems clear that specific older brain areas, such as the hippocampus, cerebellum, thalamus, and anterior cingulate, were conscripted for new functions as the brain evolved. However, some researchers believe that new brain areas were also created during the course of evolution. For example, Philip Tobias called the inferior parietal lobule the "most distinctive region of the human brain . . . the only 'entirely new structure' to have appeared in the evolution of the human brain." Other researchers have expressed skepticism that it is "entirely new" but acknowledge that this region is "virtually impossible to identify in nonhuman primates" and underwent "enormous expansion and differentiation . . . during the transition from the simian to the human condition."[20]

Regarding the question of why the brain of *Homo habilis* grew when it did, there is no widely accepted answer. Changes in climate and other environmental conditions, dietary changes such as increased meat eating, and social changes have all been proposed. One widely cited theory is the social brain hypothesis, proposed by Oxford University anthropologist Robin Dunbar. This is based on the observation that primates with larger brains live in larger social groups, and Dunbar therefore claimed that "primates evolved larger brains to manage their unusually complex social systems." In other words, as the earliest hominins came together to live in larger groups two million years ago, their brains grew larger to accommodate the more complex social relationships necessitated by the larger groups. The cause-and-effect relationship in Dunbar's theory, however, remains debatable. Larger brains would confer many evolutionary advantages in addition to managing social complexity. For example, larger visual and olfactory systems would make hominins more capable of detecting danger, and a larger memory system would help *Homo habilis* remember the location of food sources. Perhaps hominin brains grew larger for unrelated reasons, and then the larger brains allowed the hominins to manage social complexity and thus live in larger groups. The brain size conundrum is well summarized by science writer Michael Balter: "For now, just how the human brain got so big remains a puzzle. Fortunately, natural selection has already made it big enough that this is one mystery we might someday solve."[21]

In summary, approximately two million years ago some hominins living in eastern Africa began developing larger brains and becoming significantly smarter. Given the fact that hominin brains had grown very little in size during the preceding four million years, this development was unexpected and remains unexplained. The brain areas that increased disproportionately included specific parts of the frontal and parietal lobes; these areas have been linked to intelligence using modern neuroimaging techniques. This was the first of the five major cognitive advances that would ultimately produce modern *Homo sapiens* and the gods who followed. Although *Homo habilis* had become smarter than other hominins living at that time, they were not aware that they were smarter. That would come next.

2

HOMO ERECTUS

An Aware Self

The evolution of consciousness can scarcely be matched as a momentous event in the history of life.

—STEPHEN JAY GOULD, *ONTOGONY AND PHYLOGENY*, 1977

Homo habilis would merely be the starting gun for the beginning of the human race. Its modestly increased brain capacity made it smarter, so it could make tools, use the tools to make other tools, and store tools for future use. The frontal-parietal brain network associated with intelligence was developing and would continue to develop for the next two million years, making the offspring of *Homo habilis* progressively smarter. Hominins had started down the path that would ultimately lead them to, among other things, a belief in gods.

A second major cognitive leap forward for hominins was demonstrated by *Homo erectus*. These hominins first appeared approximately 1.8 million years ago and lived until 300,000 years ago, thus existing for 1.5 million years. Previously, it was thought that *Homo erectus* had descended from *Homo habilis*, but recent archeological research suggests that *Homo habilis* and *Homo erectus* lived side by side in what is now northern Kenya "for almost half a million years," making this evolutionary sequence less likely. In 2012 a third hominin species was discovered in Africa that had lived in the same area at about the same time; presumably, there were additional early species, still to be discovered, that will help to clarify the relationships between the earliest hominin species.[1]

Homo erectus was taller than *Homo habilis* and had a much bigger brain. Adults averaged five feet in height and weighed about 125 pounds. According to Andrew Shryock, an anthropologist at the University of Michigan, and Daniel Smail, a historian at Harvard University, *Homo erectus* had physical features, especially its arms and toes, that "suggest that these hominins had more or less given up climbing trees, implying a completely terrestrial lifestyle." The brains of *Homo erectus* ranged from 750 to 1,250 cubic centimeters, averaging about 1,000 cubic centimeters; therefore, their brain was about 60 percent larger than the brain of *Homo habilis*. Since the average brain capacity of modern *Homo sapiens* is about 1,350 cubic centimeters, the largest *Homo erectus* brains overlapped in size with the smallest *Homo sapiens* brains. Thus, it has been claimed with some justification that *Homo erectus* was "the first hominid [hominin] species whose anatomy and behavior justify the label human."[2]

The larger brains of *Homo erectus* led, as might be expected, to new behavioral horizons. Their stone tools, some of which have been dated to more than 1.7 million years ago, went from being crudely flaked on one side, like those made by *Homo habilis*, to being elegantly flaked on two sides. This new tool, generally referred to as a biface or handaxe even though it was really just an elegantly sharpened rock, sometimes weighed several pounds and was significantly sharper than earlier tools. According to archeologist Kenneth Feder, making a good stone handaxe "takes great skill, precision and strength. . . . Very few of my students ever developed proficiency in handaxe production."[3]

In addition to handaxes, *Homo erectus* also made what appear to be the first weapons specifically manufactured for hunting animals. These were wooden spears of up to six feet in length and sharpened at both ends. Eleven such spears found at a German site were apparently used to hunt wild horses. At sites in southern England and Spain, it appears that *Homo erectus* hunted other large mammals, including bison, deer, bears, and elephants. Such hunts would have required the cooperation of a large number of people. In addition to the sharpened wooden spears, stone-tipped spears dated to 460,000 years ago have recently been found in South Africa.[4]

Homo erectus was apparently also the first hominin to control and use fire. Exactly where and when this initially occurred is a matter of debate; there is good evidence for the controlled use of fire by 790,000 years ago, and by about 400,000 years ago the controlled use of fire had become widespread. Fire can be used for warmth, for light, for protection from predators, and for hunting by setting fires to drive animals over cliffs. One of the most important uses of fire is for cooking, which kills bacteria and parasites that may be in the food and also makes most food easier to digest. Cooking also makes meat tastier, as chimpanzees have shown by their preference for cooked meat. Fire can also be used to smoke meat, thereby allowing the meat to be stored. In one experiment, mice raised on cooked meat gained 29 percent more weight than those raised on raw meat, suggesting that cooking would have had substantial nutritional benefits for *Homo erectus*. The nutritional advantage of cooked food may have been one reason why the brain of *Homo erectus* became much larger than its ancestors. Cooking would have also promoted social interaction when early hominins gathered around campfires to share food.[5]

The larger brains of *Homo erectus* enabled them to extend their horizons not only behaviorally but also geographically. Prior to 1.7 million years ago, there is no evidence that any hominin had left the African continent. Between 1.7 million and 700,000 years ago, in what was a remarkable migration, *Homo erectus* spread halfway around the world, from present-day Spain, France, Germany, Italy, England, Israel, and Georgia to Vietnam, China, and Indonesia. In the latter two countries, *Homo erectus* fossils were initially known as "Peking Man" and "Java Man." The fact that *Homo erectus* could migrate thousands of miles and successfully survive in such varied climates testifies to this hominin's ability to adapt and cooperate in group endeavors. Since many of the areas where *Homo erectus* settled were colder than Africa, the use of animal skins as clothing and the control of fire would have been imperative. Cooperative living is also suggested by archeological remains in natural shelters, such as caves, and in the building of artificial shelters. Cooperative hunting and cooperative living both would have required some form of communication, but the degree of language development that existed at that time remains the subject of spirited debate.[6]

SELF-AWARENESS

Cognitively and behaviorally, the achievements of *Homo erectus* were remarkable. In the first four million years after separating from other primates, early hominins had only managed to make crude stone tools. Then, in the next million years, they made sophisticated handaxes, carved wooden spears with which they hunted large mammals, controlled the use of fire, and migrated from Africa to settle from England to Indonesia. Evidence of cooperative living and hunting also suggests fundamental changes in hominin relationships. As Canadian psychologist Merlin Donald noted, "With this species, a major threshold had been crossed in human evolution."[7]

What is the possible explanation for these extraordinary changes in behavior? Since the size of hominin brains was significantly increasing at the same time these behavioral changes were occurring, it is reasonable to assume the two developments were related. Clearly *Homo erectus* was much more intelligent than *Homo habilis*, but can intelligence alone explain the behavioral changes? The changes in interpersonal relationships exhibited by *Homo erectus*, as exemplified by cooperative hunting and living, suggest that something more had occurred.

A logical place to look for evidence of what might have occurred is in child development. As noted previously, it is generally accepted that the sequence in which children acquire cognitive capacities roughly parallels the sequence in which these capacities developed in human evolution. Human infants acquire increasing motor skills and intelligence for their first two years, at which time they acquire a major cognitive skill—self-awareness. Prior to that age, a child has minimal self-awareness and will often react to its image in a mirror as if it is the image of another child, trying to touch it or crawling around the mirror to find the other child.[8]

The classic experiment to demonstrate the development of self-awareness in children was done by Beulah Amsterdam at the University of North Carolina in the mid-1960s as part of her psychology dissertation. She put each of eighty-eight children, ages three months to twenty-four months, in front of a mirror, then pointed to the mirror and said, "See, who's that?" Each child had had a red mark placed on its nose to facilitate self-recognition, and self-recognition was assumed to occur if the child touched its nose or examined it in the mirror. No child younger than eighteen months showed self-recognition, and very few between the ages of eighteen and twenty months did. However, two-thirds of the children between twenty and twenty-four months showed self-recognition. This is also the developmental stage at which children start using personal pronouns such as "me" and "mine" and speak of themselves, as in "I throw ball." These are indicators of emerging self-awareness.[9]

It should be emphasized that the development of self-awareness in children is a gradual process. It develops in a series of stages and in the earliest stages may fluctuate from week to week. Its development is not dependent on achieving a specific chronological age but rather on achieving a critical level of brain development, which may vary widely among children. This is illustrated by the fact that most children with autism or Down syndrome develop mirror self-recognition but do so at a later age than other children. Similarly, it would be expected that self-awareness in *Homo erectus* would have developed slowly and would have fluctuated in its early stages.[10]

What precisely is self-awareness? Bud Craig, a neuroanatomist at Arizona State University, defined self-awareness as "knowing that I exist" and "the feeling that 'I am.'" Others have called it "the sense of one's own being," "the ability to become the object of your own attention," "the material me," and "the sentient self." Craig also noted that "an organism must be able to experience its own existence as a sentient being before it can experience the existence and salience of anything else in the environment." Evolutionarily, self-awareness probably developed to provide "updated maps of the body's state . . . necessary for the brain to regulate life," and it would have been advantageous insofar as it allowed hominins to integrate their physical and mental states. Self-awareness is also a prerequisite for most higher thought processes; without an "I" there can be no "you." As Gordon Gallup, a psychologist at the University of Albany, correctly noted, Descartes's dictum should be revised from "I think, therefore I am" to "I am, therefore I think."[11]

How might self-awareness have been beneficial to *Homo erectus*? By developing an awareness of self, *Homo erectus* would have developed a crude awareness of others and thus been able to initiate simple cooperative endeavors. Such an awareness of others would probably not have included a detailed understanding of others' thinking, such as will be described in the following chapter under "theory of mind." Rather, the awareness would have been more like that found in animals that hunt jointly, such as wolves, lions, baboons, or chimpanzees, or in three-year-old children playing in a sandbox. They are aware of one another without necessarily understanding what the other is thinking. And they are able to cooperate in simple joint undertakings, such as moving all the sand from the sandbox to the grass. Using their self-awareness, *Homo erectus* could have similarly carried out some cooperative tasks, such as keeping a fire going all night or hunting cooperatively. It is, in fact, difficult to imagine *Homo erectus* having migrated across the world and surviving, often in cold climates, for hundreds of thousands of years without having been self-aware.

Self-awareness seems so natural to us that it is difficult to imagine hominins who did not possess it. However, some people with brain dysfunction never develop it, while others develop it but then later lose it. Among those who never develop it are some children with congenital rubella or other forms of severe retardation; in one study, many of the severely retarded children were unable to recognize themselves in a mirror at any age, even if given practice sessions.[12]

Among adults, self-awareness may become impaired with some brain diseases. Some individuals with schizophrenia, for example, have an impaired awareness of self called depersonalization. Such patients have been noted to make statements such as the following: "I'm here but not here"; "I am almost nonexistent"; "I have no consciousness"; and "My feeling of consciousness is fragmented." In some individuals with Alzheimer's disease or other forms of dementia, self-awareness may be lost entirely. In one study, seven of 22 patients with moderately severe, and all six patients with very severe, Alzheimer's disease had lost their ability to recognize themselves in mirrors. Such individuals may even "converse with the person in the mirror and might try to open the door to which the mirror was attached to invite the person in." In one study, a woman with brain atrophy believed that another woman "who was identical to her in appearance, age, background, education, and so on" lived in her house. She regularly talked to the other woman in the mirror. In another case, a woman with brain atrophy who believed another woman was living in her house "had on occasion thrown a bucket of water and other solid objects at her mirror image to try and persuade it to leave the house." Such cases illustrate the importance of normal brain functioning for this critical cognitive skill.[13]

Despite its importance in human cognitive development, self-awareness is not unique to humans. At about the same time that Beulah Amsterdam was using mirrors to assess self-awareness in children, Gordon Gallup was assessing it in various primates. This idea had also occurred to Charles Darwin, who, "while visiting a zoo, . . . held a mirror up to an orangutan and carefully observed the ape's reaction, which made a series of facial expressions." The majority of Gallop's chimpanzees learned to use a mirror to explore body areas they could not see without the mirror, such as their teeth, ears, and anogenital region. Gallup also used lipstick or marking pens to put red marks on the animals' faces and ears, and some chimpanzees responded by touching those areas. By contrast, at least thirteen species of monkeys tested have shown no signs of self-recognition. Gallup noted this "decisive difference between monkeys and chimps" and concluded that "the capacity for self-recognition may not extend below man and the great apes."[14]

Since Gallup's initial experiments, self-recognition has been demonstrated in chimpanzees on multiple occasions, as well as in bonobos, in orangutans, and, only rarely, in gorillas. Other indications of self-awareness have also been noted in these animals. For example, an orangutan raised by humans and taught to use sign language spontaneously referred to itself as "me." Chimpanzees have also learned to recognize themselves in photographs. One chimpanzee who had been raised by humans indicated that it believed it was also a human by assigning its own picture to the human category. When it later was confronted by other live chimpanzees, it referred to them in sign language as "black bugs."[15]

The demonstration of mirror self-recognition in chimpanzees and other higher primates raises the question of whether it exists in any other animals. Attempts to demonstrate it in multiple species of fish and birds have all failed except for one experiment with a magpie, a member of the crow family. Among mammals, cats and dogs do not appear to be capable of mirror self-recognition, but elephants, dolphins, and some whales do. A study of three Asian elephants demonstrated clear self-recognition; one elephant even used its trunk to explore a white mark painted on its forehead. Asian elephants are known to be unusually intelligent and can be taught to respond to over one hundred different commands. Through the use of underwater mirrors, dolphins have also been observed to explore parts of their bodies that had been marked. The researchers who did the dolphin study concluded that "the emergence of self-recognition is not a by-product of factors specific to great apes and humans but instead may be attributable to more general characteristics such as a high degree of encephalization [brain development] and cognitive ability" found in all of the big-brained animals.[16]

The experiments with mirror self-recognition in primates and other animals illustrate several points. First, all of the animals except humans quickly lose interest in mirror self-recognition. The bottlenose dolphins, for example, were said to show great interest at the beginning, "but like the chimpanzees and unlike young (as well as older) humans, they rapidly lost interest in the procedure." The elephants also "lost interest quickly." Second, the experiments confirm the observations in children that self-awareness is strongly age-dependent and that there is significant variation among individuals. Some mature chimpanzees, for example, showed no interest in the mirrors. The experiments have also shown that self-recognition may be inconsistent over time; for example, one orangutan exhibited mirror self-recognition between eighteen and twenty-four months of age but not between twenty-eight and forty-two months of age. The experiments also confirm that mirror self-recognition in primates is restricted to those most closely related to humans. Numerous attempts to demonstrate it in various species of monkeys have uniformly failed; some monkeys appear to regard their mirror image as familiar, but not as self. Finally, the presence of self-awareness in great apes and hominins but not in monkeys suggests that it may have evolved separately in both species instead of having been inherited from an earlier common ancestor.[17]

Finally, the experiments demonstrate that the self-awareness existing in nonhuman primates is at a comparatively early stage of development; according to the researchers, it is said to not go "beyond incipient forms equivalent to those of 2- or 3-year-old children." As noted by evolutionary biologist Ian Tattersall of the American Museum of Natural History, "Apes exploit their images in mirrors far less comprehensively than children do. . . . Apes make no attempt to modify their images, even in ways that might make them more socially successful." This point was wittily conveyed by British physician and philosopher Raymond Tallis, who believes that the significance of chimpanzee self-recognition has been "grossly exaggerated." In regard to the use of lipstick to make marks on the chimpanzee's face, Tallis noted "that the chimpanzee did not buy the lipstick, agonize over the colour, wonder whether it would match her clothes or be in tune with current fashion, hope it would excite her partner or shock her parents, nor did she ring her friend or colour consultant for advice."[18]

THE BRAIN OF *HOMO ERECTUS*

Given the significant increase in brain size of *Homo erectus*, did self-awareness emerge at this stage of hominin development because they had a larger brain, consistent with its emergence in apes, elephants, and dolphins? It is a reasonable guess, especially since such self-awareness would have facilitated many of the behaviors exhibited by *Homo erectus*.

Self-awareness confers an ability to think about oneself, including one's needs, at more than an instinctual level as well as one's reactions to others.

What do we know about the brain of *Homo erectus*, in addition to the fact that it was much larger than its predecessors? According to those who have studied *Homo erectus* skulls, their brains had "intriguing similarities to the modern human brain." Specifically, "the chief exterior landmarks of brain anatomy are all there: the Rolandic and Sylvian fissures, the large temporal and frontal lobes, the expanded parietal lobe, and the enlarged cerebellum." In addition, the two sides of the brain are not equal, an indication of the lateralization of functions that would become a hallmark of human brains. Based on studies of stone tools made by *Homo erectus*, there is also a "pronounced right-hand preference of their makers."[19]

Although self-awareness undoubtedly involves many brain areas, recent human neuroimaging studies have identified three areas that appear to be critical parts of the brain's self-awareness network: the anterior cingulate, anterior insula, and inferior parietal lobule, as shown in figure 2.1. The anterior cingulate (BA 24, 32) lies in the medial prefrontal cortex. Although it is anatomically part of an older brain area, the anterior cingulate appears to have been evolutionarily remodeled to become functionally part of the much newer prefrontal cortex. The anterior cingulate has many functions. The insula (which does not have a BA number) lies directly behind the underside of the prefrontal cortex and is among the most recently evolved brain areas; some researchers have claimed that it has no equivalent area in monkeys. The inferior parietal area (BA 39, 40) is also one of the most recently evolved brain areas, as mentioned in the preceding chapter.[20]

Given the massive increase in the brain size of *Homo erectus*, it seems probable that white matter connecting tracts would also have been increasing in complexity. The superior longitudinal fasciculus, described in the previous chapter, includes connections to the anterior cingulate, insula, and inferior parietal area and thus plays an important role in self-awareness. Another connecting tract that may have become more prominent at this time is the uncinate fasciculus, which connects the insula to other frontal lobe areas and to the temporal lobe, including

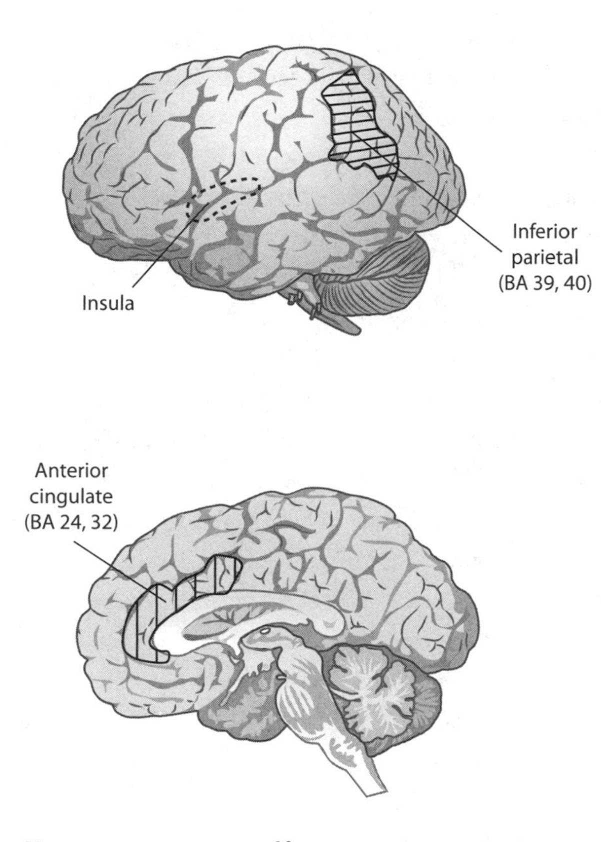

FIGURE 2.1 *Homo erectus*: an aware self.

the amygdala, which is important for the expression of emotions. Studies of the evolution of white matter connecting tracts have identified the uncinate fasciculus as one of the two most recently developed white matter tracts, consistent with its role in facilitating self-awareness.[21]

Although the anterior cingulate, insula, and inferior parietal area are known to have multiple functions, the one function they share is self-awareness. The fact that the anterior cingulate is part of this network should come as no surprise; more than a century ago, it was suggested "that the fundamental role of [the] prefrontal cortex was self-awareness or consciousness of self." Similarly, the insula is said to be "uniquely involved in interoception, the awareness of the body's internal state" and "self-awareness." Neuroimaging studies in which people's brains are monitored as they view pictures of themselves have shown activation of the anterior cingulate and the anterior insula, especially on the right side. A summary of these studies suggested that these brain areas contain "the anatomical substrate for the evolved capacity of humans to be aware of themselves." Damage to these two areas, as occurs in some cases of dementia, results in "a selective loss of self-conscious behavior and a loss of emotional awareness of self and others."[22]

The inferior parietal lobule complements the anterior cingulate and anterior insula in overseeing an awareness of self but also functions to monitor a person's body parts and their relationship to one another. This skill would have enabled *Homo erectus* to manipulate its hands more precisely in, for example, the task of making better tools and weapons. As described by Karl Zilles: "The maintenance of a spatial reference for goal-directed movements seems to be a major function of the posterior [inferior] parietal region. This function is a prerequisite for important human activities (i.e., tool use and conceptualization of actions)."[23]

Multiple human neuroimaging studies have confirmed the contribution of the inferior parietal lobule to the brain network governing bodily self-awareness. This brain area, often in conjunction with the prefrontal cortex, also becomes activated when individuals are asked to do things such as describe their own personality, recognize pictures of themselves, or imagine themselves doing various activities. For example, when healthy volunteers were shown pictures of themselves, the "medial frontal and inferior parietal lobes," among other areas, were activated. Similarly, a neuroimaging study of eight individuals who were experiencing feelings of depersonalization ("feel detached from their physical selves") reported that the inferior parietal lobule was among the areas activated. Other studies have reported activation of the anterior cingulate in similar cases.[24]

A SELF-AWARE NEURON?

Perhaps the most intriguing aspect of the development of self-awareness in hominins is the possibility that self-awareness may be the product of a particular type of brain cell that arose in the course of recent evolution. These cells are commonly referred to as von Economo neurons, or VENs, after an Austrian neurologist, Constantin von Economo, who described them in 1926. VENs are approximately four times larger than regular neurons, have a distinctive spindle shape, and are sometimes referred to as "spindle neurons." They appear shortly before birth in humans, then slowly stabilize in number during the first four years of life, eventually constituting 1 or 2 percent of the total neurons in the brain areas in which they are found. Thus, the VENs are thought to be "a phylogenetically recent specialization in human evolution."[25]

The distribution of VENS in the human brain and in other animals corresponds remarkably closely to the brain areas associated with self-awareness. VENs have even been called "the neurons which make us human." Thus, in humans, VENs have been found primarily in the anterior cingulate and anterior insula. They have also been reported to occur in much smaller numbers in the lateral prefrontal cortex but not in five other brain areas that have been examined. Among primates, VENs have been found in smaller numbers than in humans in the brains of bonobos, chimpanzees, gorillas, and orangutans, all of which have demonstrated mirror self-awareness. They have also been found in macaque monkeys, which have not demonstrated self-awareness, but not in 23 other species of monkeys, which also have not. Among nonprimate species, VENs have been described in the brains of elephants and dolphins, both of which have shown mirror self-awareness, and in whales, which have not been tested for self-awareness, but not in 30 other nonprimate species.[26]

Additional support for the importance of VENs in facilitating self-awareness comes from studies of frontotemporal dementia, a disease that begins in a person's fifties or sixties. The major symptoms of frontotemporal dementia are "a reduction in self-monitoring, self-awareness, and the ability to place the self in a social context." Thus, in the early stages of the disease individuals "behave inappropriately and with disregard for social rules, . . . have trouble comprehending other people's perspective . . . [and] have difficulty with self-awareness, showing an inability to recognize even dramatic changes in their own personality."

In contrast to people with early Alzheimer's disease, the memory of individuals with frontotemporal dementia is relatively intact. Studies that have been done on the brains of individuals with frontotemporal dementia following death have reported "severe, selective and early loss of VENs," with a 74 percent reduction of these neurons in the anterior cingulate and insula.[27]

Our understanding of VENs is still in the early stages. Since they occur mostly in mammals with large brains and in brain areas associated with self-awareness, it seems possible that VENs evolved to solve some problems related to having a large brain. One theory suggests that VENs transmit information faster than do common pyramidal neurons, making the larger brains more efficient. Thus, self-awareness, the second major cognitive leap forward, may have been one consequence of this evolutionary development.[28]

In summary, around 1.8 million years ago a new hominin, *Homo erectus*, emerged. It had a much larger brain and exhibited more sophisticated behavior than its predecessors. In addition to being more intelligent, it is likely that *Homo erectus* was also self-aware. Like Narcissus, *Homo erectus* would have been able to admire its own reflection in a still pool. With intelligence and self-awareness, *Homo erectus* had taken two of the cognitive steps necessary to become a fully modern *Homo sapiens*, able to contemplate its place in the universe and its relationship to the gods. However, it was not yet fully aware of what other hominins were thinking, or introspectively able to think about its own thinking. It was also unable to integrate time past and time present into a fully planned future. Endowed with both intelligence and self-awareness, hominins were ready to take the next cognitive step forward, one that in retrospect seems almost inevitable.

3

ARCHAIC *HOMO SAPIENS* (NEANDERTALS)

An Empathic Self

A man, to be greatly good, . . . must put himself in the place of another and of many others; the pains and pleasures of his species must become his own.

—PERCY BYSSHE SHELLEY, "A DEFENCE OF POETRY," 1821

In terms of longevity, *Homo erectus* was the most successful hominin species that ever inhabited this planet, surviving for approximately 15 times longer than our own species has so far survived. Given its success and broad geographical distribution, it is not surprising that at least 700,000 years ago, *Homo erectus* began evolving into several other hominin species, commonly grouped together and designated as Archaic *Homo sapiens*. Some members of this hominin group apparently developed a new major cognitive advance that would be essential for ultimately becoming modern *Homo sapiens* and for understanding the gods.

Depending on where they lived geographically, these hominins have been given various names, such as *Homo heidelbergensis* and *Homo neanderthalensis* (Neandertals) for those in Europe. Some specimens from Spain, dated to approximately 430,000 years ago, display features of both of these. Those in Africa are called *Homo rhodesiensis*, and recently another species has also been found there. In Indonesia a well-publicized group has been designated as *Homo floresiensis* and in Siberia as Denisovans. The Denisovans were genetically "a sister group to Neandertals," whom they apparently outnumbered and with whom they interbred. We also know that Denisovans interbred with modern *Homo sapiens* after the latter left Africa and moved eastward 60,000 years ago, since Denisovan DNA has been found in the genomes of present-day Melanesians, Australian aborigines, and natives of Papua New Guinea but not in the genomes of people living elsewhere. There were almost certainly other Archaic *Homo sapiens* species that have yet to be discovered.[1]

The best-known species of Archaic *Homo sapiens* is the Neandertals, both because they lived in Europe, where the majority of archeological research has been carried out, and because they have been immortalized by the Flintstones. They lived from approximately 230,000 to 40,000 years ago. The largest concentration of Neandertals lived in what is now southern France, but they were widely, if sparsely, distributed from Wales in the west to Uzbekistan and southern Siberia in the east. There is no evidence that Neandertals ever migrated to China or Indonesia, as their *Homo erectus* ancestors had done, or that they ever lived in Africa. Studies of Neandertal DNA suggest that their total population was relatively small.[2]

The most striking physical characteristic of the Neandertals was their large brain, which averaged 1,480 cubic centimeters, thus being larger than the average 1,350 cubic centimeters of modern humans. *Homo erectus* had achieved an average brain capacity of 1,000 cubic centimeters by 1.5 million years ago, but thereafter it increased little. However, after Neandertals evolved from *Homo erectus*, the Neandertal brain capacity increased dramatically. As Stanford University anthropologist Richard Klein noted, Archaic *Homo sapiens* "had achieved modern or near-modern brain size by 200,000 years ago."[3]

Neandertals averaged about five feet, five inches in height and weighed approximately 185 pounds, thus being significantly larger than *Homo erectus*. They had powerful upper-body musculature, and their short, stocky frames, similar to modern-day Eskimos, would have been advantageous in the cold European climate. They followed animal herds in the summer and spent winters in a home base, often a cave. Because Europe was colder than it is today, they must have made extensive use of fire and animal skins for warmth.[4]

The Neandertals were excellent hunters. They made stone tools, bone tools, and weapons that were significantly more sophisticated than those made by *Homo erectus*. In making stone tools, for example, they replaced the handaxe technique, which had been used for almost one million years, with the Levallois technique, in which flakes of predetermined size and shape were detached from a stone surface; this technique apparently developed independently in Africa and southwest Asia. Their spears, however, were "the high point of known Neandertal innovation." They were said to be "as elegantly balanced as Olympic javelins" and were used to hunt herd animals, the source of their largely protein diet. Much of the hunting was done in groups, and there is evidence of coordinated actions, such as driving herds of bison and mammoths over a cliff. They also caught fish and birds.[5]

Despite having large brains and sophisticated hunting techniques, the culture of the Neandertals is widely regarded as having been remarkably static. According to University of California anthropologist Brian Fagan, "There were no innovations, just a narrow repertoire of ancient technologies that sustained them for thousands of years." They never invented the harpoon, bow and arrow, or other weapons, despite hunting large animals for almost 200,000 years. Based on their brain size alone, Neandertals should have built computers and flown to the moon. The discrepancy between their brain size and lifestyle has puzzled archeologists and was characterized by British linguist Derek Bickerton as a "brain-culture mismatch. . . . The overriding impression of the technological evidence in the archeological record is one of almost unimaginable monotony."[6]

In recent years, some researchers have questioned whether the culture of the Neandertals was as static as has been traditionally portrayed. It has been claimed that Neandertals were using ochre, which can be used for body decoration, as early as 200,000 years ago. However, ochre can be used in a variety of ways, including as an insect repellant, for tanning skins, and for hafting stone tools onto wooden handles, so finding ochre does not necessarily mean that it was being used for decoration. There have also been two reports of ochre-stained marine shells, dated to 45,000 to 50,000 years ago, in caves thought to have been occupied by Neandertals in Italy and Spain. There is also evidence that Neandertals collected the wing bones of large birds such as eagles, falcons, and swans, as well as the talons of eagles. Some researchers have suggested that they did so in order to collect the feathers to be used as decorations, although the bones and talons may have also been collected to be used as some kind of tool. Finally, there is one report of crosshatched lines, at least 39,000 years old, carved on to a rock in a cave on Gibraltar thought to have been occupied by Neandertals. Such findings have reinvigorated the ongoing and unresolved debate regarding the cognitive abilities of Neandertals.[7]

It seems certain that Neandertals did differ from their hominin predecessors in one important respect. For the first time in history, there are suggestions that some hominins exhibited caring behavior toward other members of their group. The evidence comes from caves in Spain and Iraq. In the latter, the remains of nine Neandertals, estimated to have died 60,000 to 80,000 years ago, were found. One older man showed evidence of having had severe injuries with multiple fractures that occurred many years before his death. His injuries included trauma to his right arm and leg that would have crippled him as well as a blow to his head that likely left him blind in one eye. Such a hominin would not have survived for long on his own, suggesting that fellow Neandertals provided care for him for many years. Studies have shown that other Neandertals "suffered terribly from arthritis or had lost limbs." In order to have survived, "others in the group must have shared food with them and helped them move from camp to camp, clear evidence of pity and affection."[8]

Another possible example of caring behavior among Neandertals was their practice of at least occasionally burying deceased fellow hominins. From the period between 75,000 and 35,000 years ago, at least 59 intentional Neandertal burials at 20 sites have been found, mostly in southwestern France. Most of those interred were placed in a tightly flexed position, which some archeologists have interpreted as having had a symbolic, perhaps religious, significance. However, other archeologists noted that burials of flexed bodies may "have been purely for the practical reason that it required digging a smaller pit." Some have also speculated that the Neandertal burials suggest a belief in an afterlife. However, the burial of a dead person may simply be done to protect the person's body from being eaten by hyenas, bears, or other predators. As Brian Fagan noted, the Neandertal burials may have been "a convenient way of disposing of the departed, an essential defense strategy, especially in winter, for people living in caves often frequented by carnivores." Ian Tattersall summarized the debate about Neandertal burials by suggesting that such an act represents "at the very least a strength of attachment between individuals that transcends anything seen previously: a gesture toward the buried that was far from obligatory for any but emotional reasons." We will return to the discussion of Neandertal burials in chapter 5.[9]

A THEORY OF MIND

Providing care for another person suggests that you are able to share their emotional perspective, which is to say, to empathize with them. Empathy, then, requires an ability to get into the mind of the other person, to know what they are thinking and feeling. Psychologists refer to this as mind reading, or having a theory of mind, "an understanding that the behavior of others is motivated by internal states such as thoughts, emotions, and beliefs." It is not merely being aware of the physical presence and intentions of another person; early hominins all had that ability, as do many animals, as when dogs and wolves submit to a threatening alpha male. By contrast, a theory of mind involves actually putting yourself into the other person's mind. We read the mind of others not only by listening to what they say but also by observing their facial expressions, gaze, posture, and movements. By definition, an awareness of others cannot develop until an awareness of self has first developed, since you cannot understand the thoughts and emotions of another unless you are aware of your own, which is your point of reference. As described by Nicholas Humphrey, we can "imagine what it's like to be them because we know what it's like to be ourselves."[10]

Child development studies illustrate how an awareness of others begins to develop in children at about age four and continues until age eleven or so. This awareness of others in children follows the emergence of self-awareness at about age two, as described in the preceding chapter. Chris Frith, a psychologist at University College, London, and one of the leading researchers in the field, defined theory of mind as when we believe that other people have minds like ours and "understand the behavior of these others in terms of the content of their minds: their knowledge, beliefs and desires."[11]

A standard scenario used to assess whether a theory of mind exists in children is the Sally-Anne test. Using pictures or puppets, a child is shown Sally and Anne in a room with a ball, a covered basket, and a covered box. Sally puts the ball in the basket and leaves the room. While she is gone, the child sees Anne move the ball from the covered basket to the covered box. Sally then returns to the room, and the child is asked: Where will Sally look for her ball? In order to answer correctly, the child, who knows the ball is in the box, must understand that Sally holds a false belief that the ball's location is in the basket because she did not see Anne move it to the box. This is called a first-order theory of mind; as will be described in the next chapter, theory of mind scenarios can become much more complex.

Until the age of four, almost all children say that Sally will look for her ball in the box where Anne moved it. Children up to that age are unable to distinguish what they know from what others know. Daniel Povinelli and Christopher Prince, psychologists at the University of Southwestern Louisiana, illustrated this difficulty as follows:

> For example, sit across the table from a young 3-year-old girl and show her a picture of a turtle so that it is right-side-up from her perspective, but upside-down from yours. She will readily assent that you can see

> the turtle, and indeed, if you cover your eyes, she will readily acknowledge that you can no longer see it. But try as you may, you will have an extraordinarily difficult time getting her to understand that, from your perspective, the turtle appears differently, that is, upside-down. But less than a year later, this same child will demonstrate without hesitation her understanding that although the two of you are visually connected (or attending) to the same thing, your respective mental representations of the object differ considerably.

Starting about age four, children acquire an ability to put themselves into the mind of others. They then reply that Sally will first look for the ball in the basket, because that is where she left it and believes it to be. The acquisition of a theory of mind occurs at an earlier age in children who have older siblings and also in children whose parents more frequently use expressions referring to mental states when talking with them, thus apparently helping them to develop this cognitive skill.[12]

DO ANIMALS HAVE A THEORY OF MIND?

This raises the question of whether animals other than humans have an awareness of others' thoughts. It is generally accepted that most animals do not. For example, a baby rabbit will try to hide when it sees an eagle overhead; the rabbit does so out of instinctual behavior, not because it puts itself into the eagle's mind and speculates that it may be hungry.

Elephants have been observed performing what appear to be empathetic acts toward other elephants, but it is unclear whether this represents a theory of mind. In one instance, "a male elephant was observed for several hours providing care to a dying companion by trying to force him to stand when he fell down and bringing him water to drink." In another instance, "a drowning elephant calf was saved when the matriarch and another adult female climbed into the lake, positioned themselves on each side of the calf, and, using their tusks and trunks, lifted the calf to safety on the shore."[13]

Baboons are examples of animals that apparently have some awareness of self but do not have a theory of mind. They "distinguish clearly between 'me' and 'not me' . . . [and] identify strongly with their matriline." They are able to keep track of the hierarchal status of their troop and their own relationship with its various numbers. They have sophisticated social and communications skills. However, according to researchers who have extensively studied them, baboons appear to lack awareness of the emotions or knowledge of other baboons: "Baboons' theory of mind might best be described as a vague intuition about other animals' intentions. . . . We cannot yet conclude that baboons regard other baboons—even tacitly—as intentional beings with goals, motives, likes and dislikes."[14]

Whether or not great apes have an awareness of others' thoughts has been widely debated. It is well known that chimpanzees and gorillas can deceive others. Jane Goodall and other primatologists have described multiple instances of chimpanzees intentionally misleading other chimpanzees regarding, for example, food supplies. They have also described examples of chimpanzees helping other chimpanzees who appeared to be in distress. And in an impressive display of apparent empathy, when a three-year-old boy fell unconscious into the outdoor gorilla yard at the Chicago zoo, a female gorilla, while holding her own baby, picked up the boy as well, then carried him to the enclosure's doorway so that zoo staff could easily reach him and take him to safety."[15]

But does such behavior represent a true theory of mind, or is it rather learned behavior based on past experience? For example, if I do "x," he will do "y," and I will get all the bananas. Although the question is still being debated, there appears to be a consensus among researchers that chimpanzees and perhaps other great apes have "the beginnings of the elements of a theory of mind," "the rudiments of one," or are said to be "hovering on that crucial theory-of-mind boundary." One research group summarized the findings as follows: "We feel safe in asserting that chimpanzees can understand some psychological states in others. . . . But at the same time it is clear that chimpanzees do not have a full-blown, human-like theory of mind." Another group of researchers pretended that chimpanzees had acquired a theory of mind just long enough to deny that they had one and wittily let them answer the question for themselves:

> Yes, we share with you a psychological system that is able to knit these behaviors into novel and productive strategies that serve to fulfill our goals and desires. And it is true that our emotions, mannerisms, and reactions are much like your own. We even possess a self-concept that offers us an objective perspective on our own behavior. But what ever gave you the idea that we have a theory of mind? Why do you want to believe so desperately that we are able to construct a self–other narrative like you? After all, it was your lineage, not ours, that tripled the size of its brain during the past 5 million years. It was your species, not ours, that constructed the idea that there are unobservable mental states that mediate behavior. And thus it is you . . . not us, who are in the position of reinterpreting ancient behavioral patterns in terms of mentalistic notions—notions that never even occurred to us.[16]

WHEN A THEORY OF MIND IS IMPAIRED

The ability to think about what others are thinking—a theory of mind—would have presumably provided a major evolutionary advantage to any hominin species that acquired this skill. In food acquisition, for example, a hunter with a theory of mind could think about the strategies being used by other hunters and devise novel approaches that might be more successful. A warrior with a theory of mind probably would better predict what his enemy was going to do. A trader with a theory of mind might more accurately ascertain the minimum price acceptable to the seller of goods. And in reproduction, a man or woman with a theory of mind would be more likely to successfully seduce their partner. Indeed, the art of seduction—and passing on one's genes—focuses in part on thinking about what the other person is thinking and wants.

It is very difficult for us, as modern *Homo sapiens*, to imagine what hominins were like before they acquired an awareness of others' thoughts.

Thinking about what others think, know, believe, or desire is an integral part of being human, the essence of our daily gossip and entertainment, including movies, plays, and the ubiquitous comedies and soap operas on television. An awareness of others' thoughts and feelings is also a prerequisite for empathy, for without an awareness of others' thoughts there can be no empathy. In *The Modular Brain*, neurologist Richard Restak noted that damage to the prefrontal cortex may impair an awareness of others and thereby reduce us "to an almost subhuman level of functioning . . . devoid of what I consider our most evolved mental ability: our capacity to empathize with others."[17]

There are several human conditions in which the awareness of other people's thoughts is impaired. Foremost among these is autism; individuals with autism are said to have a "special difficulty with those situations in which it is necessary to take into account what someone else knows or expects." British psychologist Simon Baron-Cohen has called this deficit in autistic children "mindblindness." Giving the Sally-Anne test to children with autism illustrates the deficit. Among normal four-year-old children, 85 percent correctly answer that Sally will look for the ball in the basket, where she left it, rather than in the box where Anne moved it. Among children with autism, however, only 20 percent respond correctly. They have difficulties putting themselves into the mind of Sally and understanding that she holds a false belief. Autism is thought to be caused by damage to several brain areas, including the prefrontal cortex.[18]

Another human condition in which awareness of others' thoughts is impaired is antisocial personality disorder; individuals with this disorder lack empathy, often commit criminal acts, and constitute the majority of individuals in jails and prisons. Neuroimaging studies of such individuals have reported abnormalities in many brain areas, including the anterior cingulate, insula, and inferior parietal area. Accidental damage to the prefrontal area can also result in an impaired awareness of others. The classic example usually cited is Phineas Gage, who in 1848 had an iron bar penetrate his frontal lobe. Before the accident, Gage was said to have been "quiet and respectful." Afterward, he was described as being insensitive to other people's feelings and as being "gross, profane, coarse, and vulgar to such a degree that his society was intolerable to decent people," although more recent information about his later life suggests that the change in his behavior was not as stark as has usually been portrayed. Intentional damage to the prefrontal cortex such as occurs in surgical lobotomies, which were once done for individuals with severe mental illness, also commonly resulted in "a decreased awareness of the feelings of other people." Such individuals were described as being "tactless, apparently neither observing nor caring about the effect of their remarks upon their hearers," and behaving in a manner "characterized by a sometimes shocking lack of inhibitions and absence of consideration toward others."[19]

THE BRAIN OF ARCHAIC *HOMO SAPIENS*

Based on their caring behavior, it seems probable that Neandertals, and perhaps other species of Archaic *Homo sapiens*, had developed a theory of mind. If that is true, how would their brains have differed from those of their predecessors? Based on studies of their skulls, it is evident that the brains of the Neandertals were not only significantly larger than those of *Homo erectus* but also shaped differently. Specifically, according to British anthropologist Christopher Stringer, Neandertal brains had "a taller brain case and expanded parietal lobe." Other researchers have confirmed that Neandertal brains exhibited "a striking advance in the parietal area."[20]

In recent years, neuroimaging studies have been carried out on human volunteers to identify brain areas that are activated by various theory of mind tasks. Subjects have been asked to respond to questions such as the following while the activity of their brains was being assessed: "A man who has just robbed a bank is running down the street and drops a glove. A policeman, unaware of the robbery but seeing the dropped glove, calls to the man to stop so that he can retrieve the lost glove. The robber then puts his hands up and confesses to the robbery. Question: Why did the robber do this?"

The results of such studies vary somewhat depending on whether the person is asked to think about another person's thoughts, beliefs, desires, or emotions, but the overall brain activation pattern is remarkably consistent, as shown in figure 3.1. It includes the temporo-parietal junction and parts of the frontal lobe (anterior cingulate, insula, frontal pole, and the medial frontal cortex).

The temporo-parietal junction (TPJ) consists of the inferior parietal lobule (BA 39, 40) and adjacent posterior superior temporal area (BA 22). Anatomically, these areas are remarkably similar, and "most authors agree that it is practically impossible to determine the boundary between the parietal and temporal cortices" at the temporo-parietal junction. The posterior portion of the superior temporal area is especially interesting, since it includes the Wernicke speech area, usually on the left side, and an extensive association cortex. Thus, this part of the brain interprets the speech of others and, in the adjacent association cortex, puts the words into a broader context of other things known about the speaker. This is the essence of reading another person's mind.[21]

The acquisition of theory of mind in evolving hominins was probably also facilitated by the continuing development of white matter connecting tracts. The uncinate fasciculus would have been part of this development, since it connects the insula and prefrontal cortex to the superior temporal area. Another important connecting tract for the development of a theory of mind would have been the arcuate fasciculus, often regarded as the fourth part of the superior longitudinal fasciculus discussed in chapter 1. The arcuate fasciculus is a major connecting pathway between the lateral prefrontal area and the superior temporal area and temporo-parietal junction as well as other parts of the temporal lobe. Comparative studies of the arcuate fasciculus in humans and chimpanzees have reported that "the organization and cortical terminations of the arcuate fasciculus were strongly modified in human evolution."[22]

Imaging studies have demonstrated the importance of the temporo-parietal junction for a theory of mind. For example, a study of twelve volunteers who were asked to think about other people in a story reported that "the right temporo-parietal junction (RTPJ) was recruited

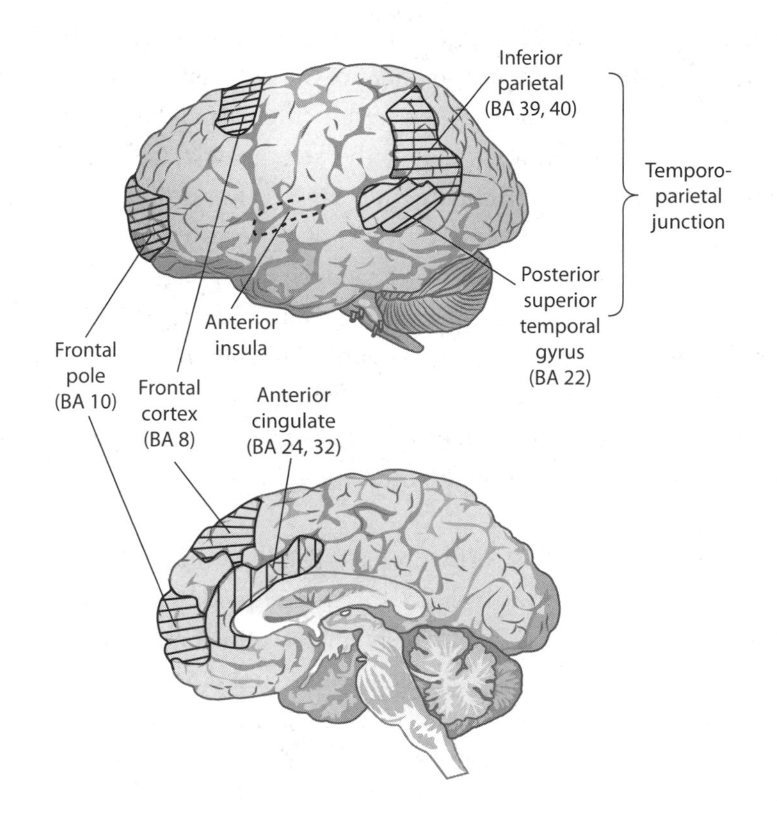

FIGURE 3.1 Archaic *Homo sapiens*: an empathetic self.

selectively for the attribution of mental states. . . . The RTPJ is highly specific to the attribution of mental states." Another study of twenty volunteers who were asked to think about individuals in family photos reported that the right TPJ was "uniquely engaged during the elaboration of ToM [theory of mind] events." A summary of such studies concluded that the right TPJ "is specifically involved in theory of mind." Especially intriguing was one study that reported that the right inferior parietal area was associated with an awareness of others, whereas the left inferior parietal area was associated with an awareness of self, suggesting a possible brain lateralization for self and other awareness.[23]

Although the temporo-parietal junction appears to be critical for reading other people's minds, it functions as part of a network with frontal lobe structures. Neuroimaging studies of individuals thinking about other people also report activities of the anterior cingulate, insula, and medial prefrontal cortex. The anterior cingulate and insula are critical brain areas for thinking about the self, as described in the last chapter, so it is not surprising that they are also important in thinking about others. The overlapping functions of these two brain areas have been noted by many researchers. The anterior cingulate has been said to be "one of several critical areas in tasks related to 'theory of mind.'" And the insula has been said to play "a fundamental role" in the mechanism by which "we understand what others feel." For example, in a study in which subjects watched videos of other people apparently experiencing injuries, both the anterior cingulate and insula were activated.[24]

Additional frontal lobe structures are also involved in the theory of mind brain network. This includes the frontal pole (BA 10), discussed in chapter 1 as being critical for perception, information processing, social cognition, the processing of emotions, and other functions. It is noteworthy that the skull of *Homo floresiensis*, the dimunitive hominin found in Indonesia and thought by many researchers to be another species of Archaic *homo sapiens*, also had an enlarged frontal polar region, suggesting that it may also have developed a theory of mind. Another prefrontal area connected to the theory of mind network is the frontal cortex (BA 8).[25]

There is one other neuroanatomical aspect of the theory of mind that is of interest. In 1996 it was reported that monkeys have neurons in their brains that fire whenever the monkey performs a goal-directed action but also fire whenever the monkey observes a similar action in another monkey. These neurons have been called mirror neurons. Neuroimaging studies have suggested that humans also possess a diffuse network of mirror neurons, widely spread in the cortex, including the insula and inferior parietal areas. Since these neurons are affected by the behavior of others, it has been speculated that "this mirror neuron mechanism may be part of, or be a precursor to a more general mind-reading ability." For example, it has been proposed that "it is thanks to mirror neurons . . . that we flinch at the sight of someone else hit suddenly by a fist or a ball and wince while reading a gruesome account of torture." Although the mirror neurons present an intriguing model for possibly understanding the neurological basis for a theory of mind, it is premature to draw any conclusions, especially since monkeys possess these neurons but have never been shown to have any awareness of others' thoughts.[26]

THEORY OF MIND AND BELIEF IN GODS

It has been recognized by several researchers that the acquisition of a theory of mind is a necessary precondition for a belief in gods. For example, in *The Belief Instinct*, Jesse Bering, a psychologist at Queen's University in Belfast, detailed how a theory of mind leads to assumptions about gods. He noted that it is only after we acquire a theory of mind that we can imagine the mind of a god. We assume, of course, that the god also has a theory of mind and thus the god can imagine what we mortals are thinking. As Bering summarized it: "God was born of the theory of mind." The requirement that hominins must have acquired a theory of mind in order to believe in a god is also stressed in recent books by University of British Columbia psychologist Ara Norenzayan (*Big Gods*) and Oxford University biologist Dominic Johnson (*God Is Watching You*), briefly summarized in chapter 8.[27]

Creating gods and attributing to them a theory of mind lead to several possible benefits. Most important, it leads to the belief that the gods can read our minds and know what we are thinking. In studies of many religions, the gods are "envisioned as possessing a deep knowing of people as unique individuals—of their 'hearts and souls.'" According to Bering, this led "our ancestors to feel and behave as though their actions were being observed, tallied, judged by a supernatural audience"—in short, to greater social order. Another benefit of creating gods with a theory of mind is that such gods are useful for explaining the unknown features of life, such as lightning being the gods showing anger, and disease being a retribution from the gods.[28]

Will Gervais, a psychologist at the University of Kentucky, developed a similar thesis regarding the importance of a theory of mind. He asserted that "the same abilities that allow people to represent and reason about each other's minds may also allow people to represent and reason about supernatural minds. . . . Therefore, mind perception is absolutely basic to religious cognition. . . . Mind perception may be the cognitive basis for belief in gods." Since this is true, reasoned Gervais, individuals who have difficulty understanding other people's minds should also hold less strong beliefs about gods. As noted earlier, individuals with autism have been said to have some impairment in their theory of mind—"mindblindness." Gervais cited studies that have reported "a modest but reliable inverse association between the autistic spectrum and belief in God." In one such study, adolescents with autism, compared to controls, were only 11 percent as likely "to strongly endorse belief in God." Such findings support the association between theory of mind and theistic beliefs.[29]

Since theory of mind is related to thinking about gods, Gervais also reasoned that there should be some overlap in the brain regions activated by both thought processes. This has been tested in a series of experiments by Dimitrios Kapogiannis and colleagues at the National Institutes of Health in Bethesda, Maryland. They did neuroimaging on subjects "with varying degrees of self-reported religiosity" to assess the brain areas activated when subjects responded to questions about God's level of involvement, God's level of anger, and religious doctrine. There was a modest overlap on the first and third questions with brain areas activated by theory of mind experiments, and the authors concluded that "religious belief engages well-known brain networks performing abstract semantic processing, imagery, and intent-related and emotional ToM [theory of mind]."[30]

It is very unlikely, however, that Neandertal hominins believed in gods. Although they apparently had acquired a theory of mind, they had not yet acquired a second-order theory of mind that would allow them to think about what god was thinking about them. Nor had they acquired an ability to fully project themselves into the past and future and to use their past experiences to plan the future. In short, they were not yet cognitively mature enough to create and honor the gods. We will return to this question in the following chapters.

Thus, by about 200,000 years ago, Neandertal hominins had brains bigger than those of modern *Homo sapiens*. They were intelligent and apparently had acquired both an awareness of self and an awareness of others. The combination of these abilities would have provided Neandertals with a significant evolutionary advantage in food acquisition, warfare, trading, and reproduction, since they would have had an ability to think about and predict others' actions.

However, they apparently still lacked an introspective ability to think about their own thoughts as well as an ability to use the detailed past and present to plan the future.

By 100,000 years ago, hominins had been separated from their primate ancestors for approximately 5.9 million years, 99 percent of the duration from the time of separation to the present. What were the odds, in the remaining 100,000 years, that hominins would build monuments such as Angkor Wat and Chartres Cathedral to honor gods, write *Macbeth* and *Messiah*, and fly to the moon? Something remarkable was about to happen.

4

EARLY *HOMO SAPIENS*

An Introspective Self

Since you have seen the dust, see the Wind; since you have seen the foam, see the Ocean. . . . Come, see it, for insight is the only thing in you that avails: the rest of you is a piece of fat and flesh.

—RŪMĪ (1207–1273), "THE GRIEF OF THE DEAD," *MATHNAWĪ*, BOOK 4

One hundred thousand years ago, multiple species of Archaic *Homo sapiens* lived in small groups throughout Africa, Southeast Asia, the Middle East, and Europe. Australia and the Americas were apparently still devoid of hominins. In appearance, except for having prominent brow ridges, they looked surprisingly modern. If properly dressed and carrying briefcases, they would not evoke undue attention today riding a subway in New York or London.

In behavior, however, they were remarkably primitive and lived pretty much as their ancestors had lived for over a million years. They controlled fire, made tools and weapons, hunted large animals, migrated long distances, and had at least rudimentary methods of vocal communication. At least one species of Archaic *Homo sapiens*—the Neandertals—had the brain capacity of modern humans, and in fact had had it for

100,000 years, yet they had relatively little to show for it. All Archaic *Homo sapiens* were probably self-aware and able to recognize themselves in a pool of still water, since they had evolved from *Homo erectus*. And the Neandertals had also apparently developed a theory of mind, the ability to think about what others were thinking; other species of Archaic *Homo sapiens* may have done so as well.

However, none of these hominins was apparently yet able to think about themselves thinking about themselves or to fully place themselves in time past and time future with what would be called an autobiographical memory. They would be completely perplexed by modern human interactions, and if you had asked them about their gods, they would probably not have had the vaguest idea what you were talking about.

THE FIRST SPARKS

The earliest definitive archeological evidence for unique behaviors we associate with modern humans comes from individuals who lived in caves and rock shelters in the Middle East and on the southern tip of Africa beginning about 100,000 years ago. The hominins in the Middle East had apparently migrated from Africa and were living in the same region where Neandertals were living at that time, and genetic studies suggest that the two groups interacted. Since there is no evidence that these African travelers survived over time or spread further, it is assumed they eventually died out or returned to Africa. However, they left behind some pierced shell beads and red ochre pigment that have been dated to 100,000 to 115,000 years ago; these may be the first known examples of self-decoration.[1]

The evidence from the caves in South Africa is more definitive and is dated to between 75,000 and 100,000 years ago. It includes highly sophisticated stone tools and weapons, produced by heating the cutting edge in a fire and then pressure flaking it to make the tip "thinner, narrower and sharper." Some of the stones chosen for use came from sites

20 miles or more distant from where the tools and weapons were being made. Cambridge University archeologist Paul Mellars claims that these stone tools and weapons from South Africa were as good as those made at European sites 50,000 years later. Also impressive was the finding of "28 shaped and polished bone tools," the first appearance of bone being used as a tool or weapon, since the bone tools used by the Neandertals appeared later. Bone tools have also been found in the Congo and dated to at least 75,000 years ago. There is also "circumstantial evidence" in South Africa for the use of snares for trapping small animals and bows and arrows for hunting, the first such appearance in history. By 65,000 years ago, the use of bow and arrow technology is more definitive.[2]

The people who lived in the South African caves and rock shelters ate a varied diet that included seafood and local game. They also lived a reasonably settled existence, using bedding made from various grasses and plants, including some plants that have "chemicals that have insecticidal and larvicidal properties against, for example, mosquitos." Such plants may also have been used by these people as herb medicines.[3]

Of great interest was the finding in these caves of seashells, dated to 77,0000 years ago, that were covered with red ochre and appear to have been deliberately perforated, thus allowing them to be strung together as a necklace or bracelet. The use of red ochre to decorate necklaces or bracelets in South Africa is consistent with the prominent role ochre played in that culture. A 100,000-year-old "ochre-processing workshop" was recently uncovered in these same caves. As noted in the previous chapter, ochre can be used for tanning skins, as an insect repellent when applied to the skin, and for hafting stone tools onto wood handles, as well as for body decoration. Thus, it is not possible to say definitely what the ochre was being used for 100,000 years ago, but the fact that it was applied to the perforated shells suggests that it was, at least occasionally, being used for decoration. In addition to the perforated shells found in the Middle East and South Africa, others dated to 75,000 years ago or earlier have been found in Morocco and Algeria, suggesting that self-decoration was widespread. Altogether, five different kinds of shells have been identified. Among the South African shell beads, it was found that "beads from two separate [archeological] layers displayed patterns of wear distinct from one another, which suggested that they had been strung and worn differently at different times." This may be the first evidence of an evolving hominin fashion statement.[4]

Also of interest in the South African caves was the finding of 15 pieces of ochre that had been modified by scraping and grinding, then deliberately engraved with a sharp instrument. The engravings consist of straight lines forming various designs. On one, for example, "the cross hatching consists of two sets of six and eight lines partly intercepted by a longer line." Some of the engraved pieces of ochre have been dated to approximately 99,000 years ago. Speculation regarding the meaning of the engraved designs has ranged from being some kind of record, to a calendar, to a piece of art. Elsewhere in southern Africa, in what is now Botswana, "a six-meter- long rock was shaped to enhance its resemblance to the head of a snake" and was dated to 70,000 years ago. Such findings led the authors of the South Africa findings to suggest that, "at least in South Africa, *Homo sapiens* was behaviorally modern" by this time.[5]

There is also evidence that the inhabitants of the South African caves had begun to wear fitted clothing at this time. Hominins had presumably been using animal skins for warmth for thousands of years, especially those members of *Homo erectus* and Archaic *Homo sapiens* who lived in colder climates in Europe and Asia. However, there are suggestions that about 72,000 years ago modern humans began to wear more fitted clothing, even in hotter climates, and that the clothing was more tailored than simple animal skin capes. The clothing presumably consisted of fitted animal skins, since any evidence of woven cloth or bone needles did not appear for an additional 40,000 years. The evidence for the introduction of fitted clothing consists of genetic studies of human lice and the fact that the body louse diverged from the head louse about 72,000 years ago. Body lice have claws adapted to clinging to clothing, not skin, and only lay their eggs in clothing. According to researchers involved in these studies, the "ecological differentiation [of body lice] probably arose when humans adopted frequent use of clothing."[6]

In addition to these behaviors, groups of hominins who lived in Africa at about this time also began to travel widely. These were, of course, not the first hominins to leave Africa; *Homo erectus* had left more than a million years earlier, and its descendants were widely distributed from Europe to Indonesia. But the migration of early *Homo sapiens* out of Africa would be different. As science writer Carl Zimmer summarized it: "In an evolutionary flash, every major continent except for Antarctica was home to *Homo sapiens*. What had once been a minor subspecies of chimp, an exile from the forests, had taken over the world."[7]

In addition to the early *Homo sapiens* who went to the Middle East about 100,000 years ago, there may have been other early *Homo sapiens* who left Africa. The major exodus, which led to the worldwide distribution of modern *Homo sapiens*, is thought to have occurred about 60,000 years ago, although, as noted previously, recent modifications of the dating of such events have suggested that this exodus may have occurred much earlier. If the exodus was 60,000 years ago, it is not clear why they left at that time. The eruption of the Toba supervolcano in Indonesia 73,000 years ago, which is thought to have affected the world climate for hundreds of years, may have been a factor. The number of *Homo sapiens* who left Africa at that time has been variously estimated at from 1,000 to a few thousand. They probably left by crossing from what is now Ethiopia to Yemen at the mouth of the Red Sea, which, because sea levels were much lower at the time, was only a few miles wide. It is widely thought that there were also subsequent migrations, but their numbers and timing are not yet clear.[8]

It is possible to re-create the worldwide peregrinations of early *Homo sapiens* by mapping the genetic variations of the male Y chromosome and female mitochrondrial DNA in present-day *Homo sapiens*. One group followed the coastline through what is now Oman, Iran, Pakistan, and India and continued down the Malay Peninsula through Myanmar (Burma), Thailand, and Malaysia to Indonesia, which was then connected by land. Evidence for their presumed shoreline migration is underwater, since sea levels are now much higher than at that time. In following this route, early *Homo sapiens* encountered groups of Archaic *Homo sapiens* who had descended from *Homo erectus*. It is now clear that *Homo sapiens* interbred with these groups, since some

Southeast Asians today carry small amounts of Neandertal DNA, while others carry DNA from the Denisovans, as noted previously.[9]

Early *Homo sapiens* reached Indonesia by at least 50,000 years ago. If they had moved at a pace of only two miles per year, the entire 8,000-mile journey from Africa would have taken about 4,000 years. Unlike *Homo erectus*, however, early *Homo sapiens* did not end their journey in Indonesia. Instead, they built boats, probably by tying together logs and reeds, and crossed approximately 40 miles of open ocean to reach Australia, which was connected to Papua New Guinea and Tasmania at that time. Although hominins had used makeshift boats to cross rivers and narrow stretches of water for thousands of years, this was apparently the first time a more extended water crossing had been made and was indicative of the planning skills of early *Homo sapiens*. The water crossing apparently involved a significant number of people; computer simulations based on the genetics of present-day Australians "suggest that more than just a boatload or two of colonists founded the area's present aboriginal population." There is evidence for *Homo sapiens* settlement in Australia 50,000 years ago, in Papua New Guinea between 49,000 and 43,000 years ago, and on the Melanesian island of New Ireland more than 30,000 years ago.[10]

At the same time that some groups of early *Homo sapiens* were going east, other groups turned northward, migrating initially to Russia and then going west across Europe or east across Siberia. *Homo sapiens* bones in western Siberia have been dated to 45,000 years ago. At a site on the Don River, south of Moscow, evidence of settlement has also been dated to between 45,000 and 42,000 years ago; this site included not only sophisticated stone tools but also bone points, a carved piece of ivory, and perforated shells, presumably used as personal ornaments. There is also evidence of early *Homo sapiens* having reached Romania, Italy, and England between 40,000 and 45,000 years ago.[11]

The speed with which early *Homo sapiens* spread across the earth was impressive. Even more impressive, however, was the speed with which they displaced the other hominin groups. As Zimmer noted, "When *Homo sapiens* arrived on a territory of *Homo erectus* . . . these other humans disappeared." Even the Neandertals, who had existed for 200,000 years and had many skills, had disappeared by about 40,000 years ago, the last remnants having apparently been pushed off the European continent onto the island of Gibraltar. Studies have shown that *Homo sapiens* increased in numbers much faster than the Neandertals, who clearly were no match for their new neighbors. As Zimmer summarized it, the *Homo sapiens* success in reproduction was part of a "spiraling pressure as clever individuals relentlessly selected for yet more cleverness in their companions."[12]

AN INTROSPECTIVE SELF

Sophisticated tools, pierced shells, fitted clothing, engravings on ochre, rocks shaped to resemble animals, boat travel in open seas—a new kind of hominin had clearly emerged. The behavior of these hominins was so at variance with the behavior of their predecessors that we designate this group as *Homo sapiens*, "wise man." And we assume that such individuals must have made some kind of major cognitive leap forward. What might that have been?

Wearing shell jewelry, decorating one's body, wearing fitted clothing all suggest that early *Homo sapiens* had become aware of what others were thinking about them. Self-adornment can be a means for advertising one's family relationships, social class, group allegiance, or sexual availability and is meant to send a message to observers. Self-adornment has been used by *Homo sapiens* in every known culture, often involving extraordinary investments of time and resources, as names such as Gucci and Cartier can attest. At the heart of self-adornment is one *Homo sapiens* thinking about what another *Homo sapiens* is thinking about him or her. This is the introspective self.

Does child development provide any clues regarding this cognitive advance? As we have seen, at about age two, children develop self-awareness, assessed by mirror recognition, and it seems likely that hominins acquired similar self-awareness beginning about 1.8 million years ago. At about age four, children begin to develop an awareness of others' thoughts, as demonstrated by the Sally-Anne test, and it is possible that at least some hominins also acquired this skill beginning about 200,000 years ago. The next major cognitive skill is acquired by children beginning at about age six and is commonly referred to as a second-order theory of mind.

What is meant by a second-order theory of mind? In the Sally-Anne test, Anne moved the ball from the basket to the box after Sally had left the room. Sally thus believed that the ball was in the basket, because she didn't see it moved, and Anne believed that Sally believed that ball was in the basket. This is a first-order theory of mind, an awareness of others' thoughts.

However, the situation changes if Sally, unbeknown to Anne, was looking in the window and saw Anne move the ball from the basket to the box. In the Sally-Anne test, the child is asked: "Where will Anne think that Sally is going to look for the ball?" This is a second-order theory of mind test, because it involves thinking about what one person thinks another person is thinking. In this case, the child has to understand that Anne will think that Sally thinks the ball is in the basket because Anne did not see Sally looking in the window when Anne put the ball in the box. Most children do not begin to acquire this cognitive skill until approximately age six.[13]

It is also possible to test children for even higher-order theories of mind. In the earlier scenario, for example, what would be the situation if, unbeknown to Sally, Anne had noticed Sally looking in the window as Anne moved the ball from the basket to the box? Anne would thus believe that Sally would believe that the ball was in the box, because she saw Anne move it there; however, since Sally did not see Anne looking at her as Sally looked in the window, Sally would think that Anne believes that Sally thinks the ball is in the basket. Some theory of mind scenarios become even more complicated by adding misinformation given by one of the participants to the other.

According to researchers in this field, a first-order theory of mind describes simple human interactions of how one person thinks another person thinks, but "it cannot entirely capture social interaction." Most social discourse involves "an interaction of minds which can be properly understood only when one takes into account what people think about other people's thoughts (second order beliefs) and even what people think that others think about their thoughts, etc. (higher order beliefs)." This is the core of most complex social interactions.[14]

The acquisition of a second-order theory of mind requires the person to view the self as an object. It is not merely looking in a mirror and recognizing the self but rather being able to think about what you look like to other people, how they see you, and what you think about how they see you. It includes being able to think about yourself thinking about yourself. It is, in short, the introspective self. The fact that early *Homo sapiens* were apparently decorating themselves and wearing fitted clothing suggests that they were thinking about themselves and how they appeared to others. Thus, this may have been the first time in hominin history that a male thought his bearskin did not look good on him and a female thought her shell necklace improved her appearance. If so, it would have marked the birth of the consumer economy.

The evolution of an introspective self would have provided early *Homo sapiens* with major cognitive advantages over other hominins, especially in social interactions and being able to predict others' behavior. It would have greatly facilitated group endeavors by *Homo sapiens*, such as group hunting, and put *Homo sapiens* at a significant advantage in warfare against other hominins who did not possess this cognitive skill. Nicholas Humphrey characterized this as having an inner eye: "Imagine that at some time in history a new kind of sense organ evolves, the inner eye whose field of view is not the outside world but the brain itself. . . . allowing him by a kind of magical translation to see his own brain-states as conscious states of mind." British sociologist Zygmunt Bauman described it as follows: "Unlike other animals, we not only know; we know that we know. We are aware of being aware, conscious of 'having' consciousness, of being conscious. Our knowledge is itself an object of knowledge: we can gaze at our thoughts 'the same way' we look at our hands and feet and at the 'things' which surround our bodies not being part of them." In humans, this ability is marvelously reflective.

Like opposing mirrors, we can contemplate ourselves, and contemplate others thinking about us, and contemplate ourselves thinking about others thinking about us, ad infinitum.[15]

Some scholars have characterized the evolution of the introspective self as the defining moment in the development of human cognition. Theodosius Dobzhansky, a geneticist at Rockefeller University, noted that man alone "has the ability to objectify himself, to stand apart from himself, as it were, and to consider the kind of being he is." This ability is "an evolutionary novelty . . . one of the fundamental, possibly the most fundamental, characteristic of the human species." Sir John Eccles, winner of a Nobel Prize, said the development of introspection was "the most extraordinary event in the world of our experience . . . the coming to be of each of us as a unique self-conscious being." Pierre Teilhard de Chardin, a French paleontologist and Jesuit priest, described it as the "hominisation of *Homo sapiens*, a consciousness to turn in upon itself, to take possession of itself as of an object . . . no longer merely to know, but to know oneself; no longer merely to know, but to know that one knows": "We must not lose sight of that line crimsoned by the dawn. After thousands of years rising below the horizon, a flame bursts forth at a strictly localized point. Thought is born." In Christian theology, the emergence of an introspective self is symbolized by the Genesis story of Adam and Eve, who eat fruit from the forbidden tree in the Garden of Eden and, for the first time, become aware of themselves and their nakedness.[16]

The introspective self would appear to be unique to humans. We sometimes wonder what cats and dogs think about themselves, but they do not think about themselves, because they do not have the necessary cognitive components. Even chimpanzees, who can recognize themselves in a mirror, have never been observed decorating themselves. And they certainly show no concern about what humans think about them, as anyone can testify who has taken a small child to the zoo and, as chimpanzees mated unmindful of their observers, tried to think how to answer their child's inevitable question: "What are they doing?"

THE INTROSPECTIVE SELF AND LANGUAGE

Is the evolution of the introspective self possibly related to the development of modern language? The origin of language is one of the most spiritedly debated issues in science. The arguments begin with the definition of language itself. Language is not simply communication, since honeybees, dogs, whales, monkeys, and many other animals communicate, often using complex sounds and behaviors. Captive chimpanzees and bonobos have been taught to communicate using sign language and keyboards with symbols. They have achieved vocabularies of more than 2,000 words and have demonstrated an ability to string several words together. Great apes such as chimpanzees also have a larynx and nasopharynx similar, although not identical, to humans and have been taught, with difficulty, to vocalize a few human sounds, as parrots can also do. Is this language? If, as some linguists argue, language is merely vocabulary and syntax, then it is possible to say that chimpanzees and bonobos have a rudimentary form of language. Most linguists, however, view language as more than word mechanics.

Like great apes and monkeys, early hominins almost certainly communicated using a variety of sounds, facial expressions, and hand signals. Indeed, it is impossible to imagine *Homo erectus* having migrated halfway around the world without having had some effective communication skills. The coordinated actions needed for hunting in groups also require communication skills, but many animals, including wild dogs, wolves, lions, baboons, and chimpanzees, do so without having developed advanced language skills. Some researchers have argued that language was acquired early in hominin evolution and may even have been a major cause of the evolution. Prominent researchers in this group include British anthropologists Leslie Aiello and Robin Dunbar, who have claimed that "the need for large groups among our early ancestors was the driving force behind . . . the evolution of language." Among many primates, grooming one another is an important means of social bonding. According to this theory, as primate groups grew in size, it became difficult for one primate to groom an increasing number of individuals. Language thus developed as a substitute for grooming: "If conversation is basically a form of social grooming, then language allows us to groom with several individuals simultaneously." If this theory is correct, then according to Dunbar "speech (and hence language) must have been in place by the appearance of *Homo sapiens* half a million years ago, at least in some form."[17]

A related argument for the early development of language was set forth by Derek Bickerton. He proposed that *Homo erectus* spoke a "protolanguage" and that true language was the product of a single gene mutation, such as the FOXP2 gene identified among a British family with language impairments, that occurred approximately 200,000 years ago at the time Archaic *Homo sapiens* evolved. Terrence Deacon, an anthropologist at Boston University, dates the inception of language even earlier, claiming that the development of language and the evolution of the human brain were both a response to the acquisition of symbolic thinking. Such theoreticians would presumably agree with University of Michigan anthropologist Thomas Schoenemann, who said: "It is difficult to escape the conclusion that language likely played a major role in the evolution of the human brain."[18]

On the other side of the debate are those who believe that the evolution of the brain came first and language second, not the reverse. Steven Pinker, a psychologist and linguist at the Massachusetts Institute of Technology, has described language as "one of the wonders of the natural world . . . an extraordinary gift: the ability to dispatch an infinite number of precisely structured thoughts from head to head by modulating exhaled breath." Such a concept of language includes consideration of the listener as well as the speaker and the possibility of conveying abstract ideas. This definition of language would thus assume the existence of self-awareness and an awareness of others' thinking as a minimum prerequisite. Viewed in this light, the development of language would also be consistent with the development of an introspective self. English neuroscientist Richard Passingham is among those who have noted the similarities between introspection and language, "hearing ourselves think," and the fact that "our inner life consists of a running commentary"; he specifically linked "the ability to reflect on one's own thoughts to language." Other researchers have similarly tied a second-order theory of mind to "the formation of linguistic conventions."[19]

It thus seems possible that the introspective self and language as we know it developed together. As Simon Baron-Cohen noted, language "is not just a transfer of information like two fax machines by a wire; it is a series of alternating displays of behavior by sensitive, scheming, second-guessing social animals." Similarly, Wake Forest University psychologist Mark Leary claimed that "language requires not only symbolic thought but also awareness of one's own communication and awareness of others as receivers."[20]

Tying language development to human cognitive development, especially the acquisition of an introspective self, also sharpens the contrast between primate language and human language. Geoffrey Pullum, a linguist at the University of California at Santa Cruz, summarized the difference as follows: "I do not believe that there has ever been an example anywhere of a nonhuman expressing an opinion, or asking a question. Not ever. It would be wonderful if animals could say things about the world, as opposed to just signaling a direct emotional state or need. But they just don't." The reason they don't, of course, is that they lack the cognitive network necessary for thinking about themselves and others. Steven Pinker reflected this clearly, when he said, "deep down chimps don't 'get it.'" And University of Rochester anatomist George Washington Carver succinctly summarized it: "The only reason that an ape does not speak is that he has nothing to say."[21]

There is also anatomical evidence to support human language as a relatively late evolutionary acquisition, developing in tandem with development of the frontal and parietal lobes. The speech areas of monkeys and great apes, used to make complex vocal calls, are not in the recently evolved brain cortex, as those of humans are, but rather in phylogenetically older brain areas, in the limbic system and brain stem. Humans also use these older speech areas but only when, for example, we swear as we hit our finger with a hammer or when we cry or laugh.

Most human speech, by contrast, is controlled by two brain areas that have developed in the cortex relatively recently. The first is Broca's area, located in the frontal lobe; this controls verbal speech and is anatomically situated adjacent to the brain region controlling the muscles for the mouth, tongue, and larynx. The second speech area is Wernicke's area, discussed in the last chapter, which is located in the superior temporal lobe adjacent to the temporo-parietal junction; this controls speech comprehension and is anatomically part of the brain region associated with hearing. It thus appears that the brain areas associated with the development of self-awareness, an awareness of others' thinking, and an ability to think about one's own thinking overlap with brain areas associated with the development of language.[22]

Finally, there is linguistic evidence supporting human language as a relatively late evolutionary acquisition. Quentin Atkinson, a psychologist in New Zealand, analyzed 504 world languages for phonetic complexity to ascertain which were more complex (developed earlier) and which were less complex (developed more recently). Atkinson reported that the oldest languages were in Central and South Africa, with others following closely the migration pattern of *Homo sapiens* as we spread out of Africa. The ability of humans to speak a modern language and the evolution of our ability to think about ourselves thinking about ourselves thus appear to parallel each other.[23]

It seems more likely, therefore, that language was an accelerant of human evolution rather than its cause. What good is the ability to think about yourself if you cannot talk about yourself? What good is the ability to think about what others think if you cannot gossip about them? What good is it to think about what others think of you if you cannot talk to them or others about it? The acquisition of an introspective self would have been an immense impetus to language development. This was illustrated by Robin Dunbar, who "has eavesdropped on people on trains and in cafeterias, and he consistently finds that two-thirds of their conversations are about other people."[24]

The simultaneous development of an introspective self and language would also be synergistic from an evolutionary point of view. Each would independently improve genetic fitness, but the people who could both think introspectively and talk about these thoughts would be able to discuss complex behaviors and therefore be more successful in passing on their genes. Early *Homo sapiens* was thus the first hominin who had a lot to talk about. And, as Steven Mithen noted, "Once Early Humans started talking, they just couldn't stop."[25]

THE INTROSPECTIVE SELF AND THE GODS

The acquisition of the introspective self was a defining event in the cognitive development of hominins. As Teilhard de Chardin noted, it is "no longer merely to know, but to know oneself; no longer merely to know, but to know that one knows."[26] At the developmental equivalent of about two years of age, we had acquired the ability to think about ourselves; at about the developmental equivalent of four years of age, we had acquired the ability to think about other people's thoughts. Now, at the developmental equivalent of six years of age, we had acquired a second-order theory of mind, an ability to think about what another person thinks about us.

At first glance, this cognitive ability would appear to make it possible for early *Homo sapiens* to conceive of, indeed to worship, the gods. By acquiring a theory of mind, Archaic *Homo sapiens* acquired the ability to appreciate that gods also had thoughts. Then, by acquiring a second-order theory of mind—introspection—early *Homo sapiens* acquired an ability to think about the fact that the gods may be thinking about us, and what they may be thinking, and what we think about what the gods are thinking about us. In short, early *Homo sapiens* had acquired the cognitive ability to enter into a conversation with the gods, just as modern *Homo sapiens* does today.

But wait—where did the gods come from 100,000 years ago? Early *Homo sapiens* certainly had conversations with other early *Homo sapiens* regarding what they thought of one another, and what they thought about a third early *Homo sapiens* who had insulted them, and why they were no longer on speaking terms with the other person, ad finitum, just as happens today. But you cannot have such conversations about the gods or with the gods unless you have gods.

One theory regarding the origins of gods is the human tendency to anthropomorphize—attribute human agency to—inanimate things or events. Thus we assume that thunder and lightning, floods and droughts, the rising of the sun and the phases of the moon must all be caused by some superhuman or divine power. Such pattern-seeking theories have given rise to multiple theories of the origins of gods and religions, as will be discussed in chapter 8. Perhaps 100,000 years ago early *Homo sapiens* listened to the thunder, watched the lightning, and decided that there must be gods living in the sky who were watching them.

Since early *Homo sapiens* had acquired an awareness of the thoughts of others as well as the ability to think about what others were thinking, such a scenario for the creation of gods is theoretically possible. For a variety of reasons, however, it seems unlikely. First, why should an explanation for thunder and lightning require the concept of gods or other unseen spirits rather than phenomena that were more familiar to early *Homo sapiens*? Possible examples include such things as large animals who lived in the sky, or trees falling in an unseen world. Second, no religious symbols, effigies, or other artifacts have been found from this period that might have had some religious meaning. Much later, when it is known that gods did exist, such artifacts became very common. Third, an understanding of the importance of recurrent natural phenomena requires a cognitive ability to fully integrate the past and present into thoughts about the future. As will be described in the next chapter, early *Homo sapiens* apparently had not yet acquired this ability. Fourth, it may be questioned whether understanding natural phenomena is a sufficient stimulus, by itself, to elicit the creation of gods. When the gods finally emerged, they impelled some believers to build pyramids and cathedrals, spend long periods praying to the gods, forgo sexual pleasure by remaining celibate, and lay down their lives in warfare to defend their gods. Pattern-seeking theories would not seem powerful enough to elicit such personal sacrifice. For all of these reasons, it seems doubtful whether gods existed among early *Homo sapiens* 100,000 years ago.

THE BRAIN OF EARLY *HOMO SAPIENS*

In view of the impressive behavior exhibited by early *Homo sapiens*, one would expect to find equally impressive changes in their developing brain. But since the brain had already reached an average 1,350 cubic centimeters at least 100,000 years earlier, it could not grow larger, or the head of newborn babies would no longer fit through the bony outlet of a woman's birth canal. The brain changes leading to early *Homo sapiens*, therefore, did not involve an increase in brain size but rather internal changes. This is an answer to the problem posed by linguist Derek Bickerton: "Any adequate account of how our species came into existence has to explain how it was that the brain grew to at least its present size without changing the hominid [hominin] way of life in any significant manner, and then, without further increase, made possible the stunning explosion of creativity that characterized our species."[27]

The brain areas associated with an introspective self have been well studied in recent years using neuroimaging techniques. Typically in such studies, "subjects are presented with trait adjectives or sentences and are asked whether the trait or sentence applies to them"; at the same time, their brain is being scanned by a PET or functional MRI machine. A meta-analysis of 20 such studies, done between 1999 and 2009, identified four major brain clusters that are activated by introspective thinking, as seen in figure 4.1.[28]

One cluster is the anterior cingulate (BA 24, 32) and insula, areas that are also activated by an awareness of self and by an awareness of others' thinking. Thus, it would be surprising if these areas were not also activated by introspective thinking. A second brain cluster activated by introspective thinking includes parts of the prefrontal cortex, including the frontal pole (BA 10), the lateral prefrontal cortex (BA 9, 46), and the orbitofrontal region (BA 47). This is consistent with observations claiming that "self-awareness, consciousness, or self-reflectedness" is "the highest psychological attribute of the frontal lobe." Similarly, a review of studies of "social cognition," defined as including "self-reflection, person perception, and making inferences about others' thoughts,"

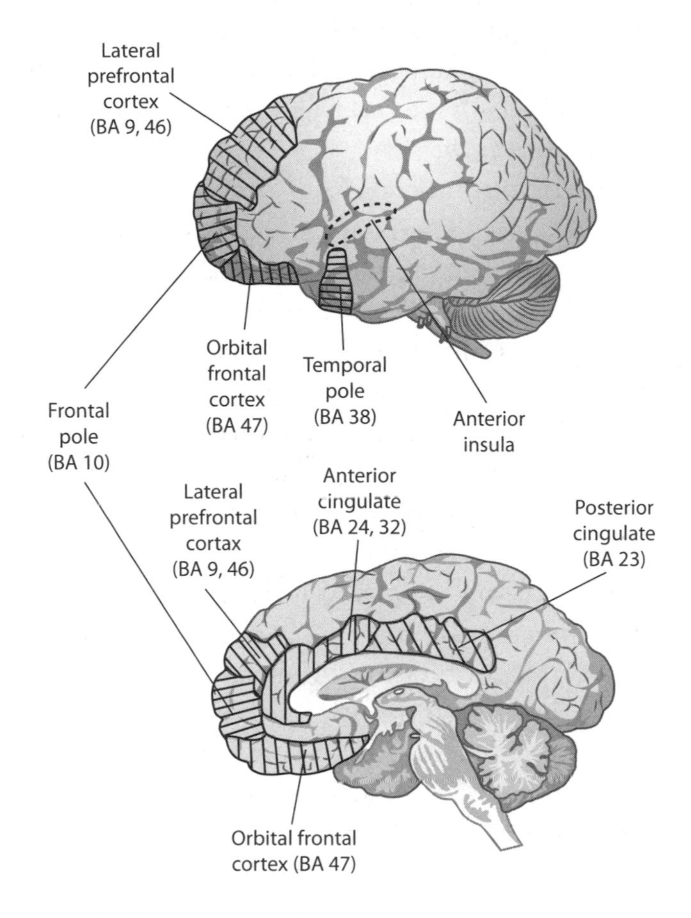

FIGURE 4.1 Early *Homo sapiens*: an introspective self.

concluded that the broadly defined medial prefrontal area has "a unique role" in social cognition.[29]

A third brain area activated by introspective thinking is the posterior cingulate (BA 23), situated in the midline behind the anterior cingulate. It is strongly connected to many parts of the prefrontal cortex as well as to the temporo-parietal junction. The posterior cingulate is said to have become activated "when subjects had to indicate whether a word or statement was self- descriptive or not." The final brain area activated by introspective thinking is the most anterior portion of the temporal lobe, called the temporal pole (BA 38). This is an older brain area that is not well understood but is known to play some role in thinking about other people's thinking. It has been shown in neuroimaging studies to be especially activated by "tasks that require one to analyse other agent's emotions, intentions or belief."[30]

At the same time that the brain areas associated with introspective thinking were developing, it seems likely that the white matter connecting tracts were also continuing to develop, especially the superior longitudinal, uncinate, and arcuate fasciculi. As described in the previous chapters, these tracts connect the anterior cingulate, insula, and prefrontal cortex anteriorally and the parietal and temporal regions posteriorally. One can easily imagine these areas becoming progressively interconnected as the white matter tracts mature, initially to make possible first-order theory of mind thinking, then second-order theory of mind thinking, then increasingly complex theory of mind assessments, and finally the ability of individuals to think about themselves thinking about themselves thinking about themselves ad infinitum.

Indeed, one of the great mysteries in human evolution is how human behavior changed so dramatically over the past 100,000 years, a brief moment in time compared to the entire span of human evolution. It seems likely that a major answer to this question will be found in the development of the white matter tracts and the cognitive skills and behaviors that became possible as the disparate brain areas became increasingly interconnected.

Thus, by 60,000 years ago, when early *Homo sapiens* are thought to have left Africa, they had apparently acquired intelligence, an awareness of self, an awareness of others' thinking, and, most remarkably, the ability to think about themselves thinking about themselves. They had acquired the cognitive skills they would need to displace all other existing hominins and to become the lords of the earth. But they had apparently not yet acquired one additional skill they would need to become truly modern *Homo sapiens*. That would come next.

5

MODERN *HOMO SAPIENS*

A Temporal Self

Time present and time past
Are both perhaps present in time future,
And time future contained in time past.

—T. S. ELIOT, *FOUR QUARTETS*, 1952

The *Homo sapiens* who exited Africa about 60,000 years ago, perhaps earlier, were remarkable creatures—intelligent, self-aware, empathic, and introspective. In the following few thousand years, they spread widely, from Australia and Papua New Guinea in the east to Europe in the west, interbreeding with, and ultimately displacing, the older hominin species as they went.

But as remarkable as these people were, you would not want your son or daughter to marry one of them. Although impressive as hominins, they were still missing the one critical cognitive skill that would be needed to become truly modern *Homo sapiens* and to worship the gods. It seems likely that the fundamental brain changes needed for the development of this skill were well underway at the time these *Homo sapiens* left Africa. Thus, this cognitive skill continued to evolve over the next several thousand years, no matter where they ultimately settled, another apparent example of parallel evolution.

Beginning approximately 40,000 years ago, several new behaviors that we associate with modern *Homo sapiens* appeared. Specifying a period, such as 40,000 years ago, is, of course, merely a milestone. Evolution is a continuous process, and hominin brains were undergoing constant change. It should also be remembered that the dating of events in the distant past, even within the past 40,000 years, is subject to at least a 10 percent error; thus, something dated to 37,000 years ago may have actually taken place earlier than something dated to 40,000 years ago.

Nevertheless, there is clear evidence that after *Homo sapiens* left Africa, the behavior of both those who left and those who remained underwent significant changes. For example, improvements were seen in the development of tools and weapons. Beginning about 49,000 years ago, there was said to be "a gradual abandonment of the technology and tool types" that had been used by early *Homo sapiens* and their predecessors. Although bone tools had occasionally been used by early *Homo sapiens* in South Africa as far back as 75,000 to 100,000 years ago, a "bone industry" emerged at this time. For the first time, animal bones, reindeer antlers, and mammoth ivory tusks became widely used as raw material for making tools and weapons. Bone, antlers, and tusks had, of course, been available for hundreds of thousands of years, but they had never previously been widely exploited as raw material. According to University of Alaska biologist Dale Guthrie, "Bone, antler and ivory are composites . . . [making them] harder and more durable than wood yet lighter and less subject to breakage than stone." What followed was the emergence of a much broader variety of tools and weapons, including "spear points, chisels, wedges, spatulas, awls, drills, needles, perforated antlers used as shaft straighteners, and, later on, harpoons and spear propellers."[1]

Needles, for example, permitted the sewing of clothing, for which there is evidence that this occurred approximately 35,000 years ago in caves in the Republic of Georgia. Wool fibers from goat hair and plant fibers from flax, some of them dyed, were apparently used to make ropes, nets, and baskets as well as clothing. The sewn clothing would have been especially useful during the Ice Age that was to follow. As Brian Fagan noted: "For the first time, women could fashion garments tailored to size for infants, children, and growing youths, as well as for adults and old people. They could also sew clothes of all kinds."[2]

Another tool that came into use at this time was the lamp. The controlled use of fire had been known for at least 500,000 years, but the use of lamps was new. Many lamps have been found in caves in France and Spain where drawings and painting were created, as will be discussed later. Most lamps were stones with a natural depression that could be filled with tallow; lichen, moss, conifers, and juniper were used for the wick. A few of the lamps that have been found have engravings on them, and one even has a handle.[3]

In addition to the tools, new and improved weapons also were developed beginning about 40,000 years ago. Spear throwers, also called atlatls, appeared about 30,000 years ago and could guide a spear faster, longer, and more accurately. They were especially effective for hunting dangerous animals, since a hunter using an atlatl could throw the spear from a safer distance.

The bow and arrow, which had been used occasionally in earlier millennia, came into widespread use by at least 20,000 years ago; it also allowed a hunter to remain at a safer distance from the prey and enabled humans to kill birds in flight. Given weapons like the atlatl and bow and arrow, modern *Homo sapiens* began to hunt animals that had previously been inaccessible, such as the dangerous Cape buffalo in South Africa and the elusive ibex in Spain. The use of fish hooks and netting also enabled *Homo sapiens* to fish in deeper waters. In Indonesia, for example, the bones of tuna and shark dated to more than 40,000 years ago provide evidence for pelagic fishing at that time.[4]

It was not merely the use of new tools and weapons that was impressive but also the speed with which they were introduced and improved. This is said to have demonstrated "an extraordinary increase in the human capacity to create and invent." In the *Prehistory of the Mind*, Steven Mithen likens this period to "a Paleolithic arms race":

> It is not simply the introduction of new tools at the start of the Upper Palaeolithic [45,000–11,000 years ago] which is important. It is how these were then constantly modified and changed. Through the Upper Palaeolithic we can see the processes of innovation and experimentation at work, resulting in a constant stream of new hunting weapons appropriate to the prevailing environmental conditions and building on the knowledge of previous generations.

The speed of such technological innovation and experimentation stands in sharp contrast to the preceding hundreds of thousands of years in which change was glacially slow, when it occurred at all.[5]

Other items that became more widespread beginning about 40,000 years ago are what appear to be memory devices. These include pieces of bone that have been engraved with a series of lines or dots, similar to the engraved ochre dating to more than 90,000 years ago found in South Africa and described in the last chapter. Alexander Marshack, a self-taught archeologist at Harvard's Peabody Museum, carried out extensive studies of these engraved bones, studying the incisions with a microscope and speculating on how they had been used. The best-known example, found in France, dates to approximately 30,000 years ago and was originally interpreted by Marshack as representing the phases of the moon during two lunar cycles. Marshack claimed the engraver of the bone "not only had an image of the waxing and waning of the moon, but he had also created an abstracted image of the continuity and periodicity of time itself." Sir John Eccles put a picture of this bone on the cover of his book *Evolution of the Brain: Creation of the Self*. Marshack referred to the bones as "notational devices," and in his book *The Roots of Civilization*, published in 1972, suggested that they represent "the evolved, time-factored, and time-factoring human capacity, the cognitive ability to think sequentially in terms of process within time and space." In more recent years, critics have charged Marshack with over-interpreting the meaning of the incised bones, even as most agree that the bones represent some sort of "external memory devices," although not necessarily for recording lunar cycles.[6]

In addition to using new tools, weapons, and memory devices, beginning about 40,000 years ago modern *Homo sapiens* exhibited increasingly varied and sophisticated forms of self-ornamentation. Randall White, an archeologist at New York University, described it as an "explosion of items of bodily adornment at the beginning of the Upper Paleolithic. . . . The emergence of technology for the manufacture of body ornaments in Europe is sudden and the technology itself is complex and full-blown from the beginning." Whereas the necklaces and bracelets found earlier in South Africa had been made with only seashells, the newer forms of self-ornamentation also used animal teeth, animal bones, antlers, ivory tusks, snail shells, bird claws, ostrich eggshells, and colorful stones to make rings, pins, and pendants in addition to necklaces and bracelets. One necklace, for example, was made from "nearly 150 perforated antler, bone and stone beads and 5 pendants, some of them incised and decorated." Some sites in the Dordogne region of France are said to have been "factories" for making beads and pendants, where "the manufacture of items of personal adornment became an industry in itself."[7]

Examples of self-ornamentation during this period are geographically widespread in Europe and the Middle East, having been found in France, Spain, the Czech Republic, Bulgaria, Lebanon, and Turkey. In the latter, necklaces made of snail shells and bird claws have been dated to 43,000 years ago. Similar examples of self-ornamentation have been found in Africa, including in Morocco, Algeria, Kenya, Tanzania, and South Africa. The use of ostrich eggshells was especially common in Africa. Everywhere there appears to have been distinct regional styles, depending on the availability of materials, with materials that were most unusual, such as luminous white or brightly colored shells, being most highly valued.[8]

Materials used for self-ornamentation also suggest the development of extensive trade networks during this period. Seashells used for necklaces in France have been found more than 100 miles from their origin. Some archeologists have suggested that exchanges of materials used for self-ornamentation may have played an important role in the development of allied groups of people: "Current theories suggest that extensive networks of marriage, friendship, and exchange grew up along chains of bead-giving and receiving." Such gifts may also "function as mnemonic devices on a par with place names and kin terms." Thus, a necklace of incised red deer teeth may represent friendship between one group and another, unrelated group that shared a cooperative hunt of red deer the previous autumn.[9]

INTENTIONAL BURIALS WITH GRAVE GOODS

One of the most dramatic new behaviors exhibited by modern *Homo sapiens* was the addition of grave goods to some intentional human burials. Hominins had been dying for six million years, but for most of that time there is no evidence that anything had been done with the bodies of deceased hominins other than leaving them on the ground to decay or to be eaten by scavengers. The first definite evidence of intentional burials occurred between 90,000 and 100,000 years ago; 11 relatively intact individuals were found buried in the floors of caves in Israel, presumably placed there by early *Homo sapiens* who had apparently migrated to southwest Asia. Between 75,000 and 35,000 years ago, multiple intentional burials were carried out by Neandertals. As noted in chapter 3, such intentional burials may represent caring behavior in these hominins, who had acquired a theory of mind, or may merely represent a method of disposing of a dead body so that it did not attract predators.[10]

Then, beginning approximately 28,000 years ago, a series of remarkable burials took place in which useful and valued items were buried with the deceased. Such items are called grave goods. The oldest such burial discovered to date was at Sungir, 120 miles northeast of Moscow. A man and two children were interred with an "astonishing material richness" that almost certainly was intended to assist them in an afterlife. The three bodies were dressed in clothing decorated with over

13,000 ivory beads; recent research has estimated that one bead would have taken up to one hour to carve. The arms of the man were adorned with 25 polished ivory bracelets, and he wore a red pendant around his neck. The adolescent male child wore a belt decorated with 25 fox teeth and had an ivory, animal-shaped pendant around his neck; beside him was a carved ivory mammoth and an ivory disc. The female child's body was accompanied by three ivory discs with intricate latticework, several ivory lances, and two antler batons, one of which was decorated with rows of drilled dots. Both children had six-foot mammoth tusk spears at their sides. In a northern climate in which obtaining sufficient food, clothing, and shelter must have occupied much time and energy, this was an extraordinary array of material goods to bury in the ground.[11]

Similar burials, dated to 27,000 years ago, have been discovered in the Czech Republic, 1,200 miles southwest of Sungir. At one site, 18 individuals were buried together, covered by mammoth bones and limestone slabs. At another site, Dolní Věstonice, a young woman and two young men were interred together, with the woman lying between the men. The positions of the bodies have elicited spirited discussion among researchers: the woman's head was turned toward one man, who was looking away, while the other man was turned toward the woman with both his hands resting on her crotch. The heads of the men were encircled with pendants of pierced arctic fox, wolf teeth, and mammoth ivory. Red ochre and mollusk shells were also abundant in the grave. The Dolní Věstonice site was occupied for at least 2,000 years by mammoth hunters who lived in houses made of wood and mammoth bone. One of the houses had been constructed using 23 tons of mammoth bones. Over 700 clay figurines have been found at the site, the world's first-known ceramics, as well as a Venus figurine dated to 26,000 years ago, some phallic carved ivory batons, and the remnants of what appears to be basketry.[12]

Another unusual burial, also dated to 27,000 years ago, was unearthed in Austria. Two newborns, thought to be twins, were covered with ochre, decorated with ivory beads, then placed in the grave beneath a large mammoth scapula that was supported by a piece of ivory tusk. The scapula thus sheltered the bodies from the dirt used to fill the grave.

Similarly, in Italy a grave of two children has been dated to this same period. They were buried with more than 1,000 perforated snail shells "arranged around the pelvis and the thighs, possibly decorating loincloths." Another double burial in Italy that may be even older included an elderly woman and an adolescent, "their heads protected by stonework"; the adolescent had four rows of snail shells around its head, and the woman had two shell bracelets. In Siberia, a young boy, wearing "an ivory diadem, a bead necklace, and a bird-shaped pendant," was buried under a stone slab 24,000 years ago.[13]

The total number of known burials from this period, such as those described earlier, is relatively small, but there are reasons for this. Since most such burials apparently took place in the open rather than in caves, it is difficult to know where to look for them. Several of the known burials have been accidentally discovered during modern construction projects. In addition, many of the earliest burials were found in central or eastern Europe, where, compared to France and Spain, much less archeological research has been carried out.[14]

Some scholars have disputed whether these burials were truly the first-known burials with grave goods. In Australia, a 40,000-year-old burial of a *Homo sapiens* had apparently been covered with red ochre, a pigment that had been carried from 120 miles away. In South Africa in what is probably an even older burial, an infant was buried with "a perforated Conus shell, presumably an ornament or amulet" that had been brought from 35 miles away. Other scholars have argued that some of the earlier Neandertal burials also included grave goods. Much of the debate centers on the definition of what constitutes grave goods. On one side of the debate are those who argue that grave goods should include such things as "ochre, implements of stone or bone, unmodified animal bones, etc." Advocates for this position point to the finding of animal bones and deer antlers in the early Israeli cave burials as evidence of intentional grave goods.

One study reported that more than one-third of intentional human burials between 75,000 and 35,000 years ago included such grave goods. On the other side of the debate are those who argue that the floors of the caves where such burials took place were probably littered with stone tools, bones, and occasional pieces of ochre and thus it would not be surprising if some of these had been inadvertently included when the grave was being filled. As Ian Tattersall noted, the bones and antlers included in the graves "are hardly very impressive" as grave goods. Such critics point out that no grave prior to 40,000 years ago included grave goods such as shells, beads, bowls, or other items that were commonly included as grave goods in later burials.[15]

An especially contentious debate has centered on a 50,000-year-old multiple Neandertal burial in Shanidar Cave in Iraq. Included in two of the graves was an abundance of flower pollen, which "suggested to its excavators that the deceased had been laid to rest on a bed of spring flowers." For many years, this burial was cited as proof that Neandertals not only buried their dead but did so with ritualistic behavior suggesting a belief in an afterlife. More recently, however, it was discovered that local rodents, called jirds, have burrows in this area and that jirds commonly store seeds and flowers in their burrows. As Richard Klein and Blake Edgar noted in *The Dawn of Human Culture*, "The jird explanation is less exciting than the human one, but it is in keeping with the total lack of ritual with other Neandertal burials, including the others at Shanidar Cave." The relationship between grave goods and a belief in an afterlife will be discussed in more detail later.[16]

THE ADVENT OF THE ARTS

Among the new behaviors exhibited by modern *Homo sapiens* beginning approximately 40,000 years ago, the advent of the arts has elicited the most popular interest. According to researchers, the advent of the arts, especially the visual arts, is of interest to modern humans because it suggests "the origin (or one of the origins) of something that is quintessentially human, something that sets us apart from animals and prehumans." The visual arts were created "at a time when human society as we know it was being born." Since writing did not begin until thousands of years later, these artistic creations are as close as we will ever come to having a record of this period.[17]

Some archeologists have questioned whether modern *Homo sapiens* was really the first hominin to create works of art. As previously noted, some Neandertal experts have claimed that Neandertals used perforated teeth and bones as pendants, made figurines out of ivory, and even modified a piece of flint to resemble a mask. The authenticity of these, however, has been disputed. Even if some of them are authentic, Paul Mellars observed, "the sheer scarcity and isolation of these objects . . . make it difficult to see this kind of symbolic expression as a real and significant component of Neandertal behavior." By contrast, the production of visual art by modern *Homo sapiens* was said to be "an orgy of artistic creation at every level, from the mere scribble to true masterpieces."[18]

The variety and abundance of visual art created during this period are impressive. Although the polychrome cave paintings are best known, artists of this period also made engravings, clay models, sculptures, figurines, and decorated objects of all kinds. Over 300 caves have been found containing art, the majority in France and Spain. Lascaux Cave alone contains 1,963 paintings and engravings, half of which depict animals, the remainder being geometric figures. The abundance of art is also remarkable, given estimates that the total population of France 22,000 years ago was only 2,000 to 3,000 people, and for all of Europe, about 10,000.[19]

The earliest examples of art from this period discovered to date are stenciled handprints from the Leang Timpuseng cave on the island of Sulawesi in Indonesia and drawings of geometric figures from El Castille cave in Spain, both dated to about 40,000 years ago. The earliest findings from the major painted caves are the extraordinary paintings in Chauvet cave, the earliest of which are dated to 36,000 years ago, and a drawing of a pig-like animal, dated to at least 35,400 years ago, in the cave on Sulawesi. The era of major cave art continued for over 20,000 years, with more recent examples being caves at Altamira (14,000 years ago) in northern Spain, and Niaux (13,000 years ago) in southwestern

France. The last cave art that has been found is in caves in Sicily at Levanzo and Addaura, both dated to about 11,000 years ago. After that, the European cave art tradition appears to have died out, at the same time that Europe was warming and the agricultural revolution was getting underway.[20]

At about the same time that the earliest cave paintings were being done, the earliest known sculptures were also being carved. In a series of caves in the Swabian Alps in southern Germany, sculpted ivory figurines of a lion, a mammoth, a bison, and a man with a lion's head and a figurine of a woman's body, all dated to 35,000 to 40,000 years ago, were found. The last was apparently used as a pendant and is a forerunner of similar female figurines carved over the next 10,000 years, found widely across central Europe and commonly referred to as Venus figurines. The most famous example, found in the Willendorf Cave in Austria, is adorned with bracelets, has an elaborate hairstyle, and was covered with red ochre. The common characteristics of these figurines are grossly exaggerated breasts, hips, and vulvas, leading researchers to assume that the figurines were associated with fertility or food surplus. Anthropologist Robin Dunbar described these figurines as "Michelin-tyre ladies," while archeologist Paul Mellars characterized them as "paleo-porn." In addition to the female figurines, many sculptures of animals, carved out of ivory, also date to this period: mammoths and lions were the subjects of choice, but horses, bears, and bison were also carved. Eight bone flutes, made from the wing bones of vultures and swans, have also been found, with the oldest dated to 42,000 years ago; these are the first known musical instruments.[21]

For many years, it was thought that there had been a progression in the European cave art, from simpler forms in the earliest centuries to more complex forms in later centuries. Such theories were proven wrong with the discovery of Chauvet Cave in 1994, which has paintings equally as sophisticated as any found at Altamira almost 20,000 years later. What does seem clear is that there was an increasing production of visual art during this period, and this production was seen everywhere modern *Homo sapiens* settled. Thus, by about 15,000 years ago in Europe, "people decorated their harpoons, spear points, spear-throwers, and other artifacts with naturalistic engravings, fine carvings of wild animals, and elaborate schematic patterns." In South Africa, ostrich shell water bottles were being incised with geometric designs. In Namibia, pictures of felines, rhinoceroses, and a giraffe-like creature were being painted on flat stones. In Australia, Brazil, and India, geometric designs and pictures of animals were decorating rock shelters. And in China, deer antlers were being incised with abstract designs.[22]

Since these visual art productions are essentially the only "written" record we have from this important period, it is worth examining them in more detail. In the caves, there are three main themes: animals, human handprints, and geometric figures. Animals are very prominent, with the vast majority of them being animals that were being hunted. Thus, an analysis of 981 cave art animals reported the following statistical breakdown: horse 28 percent, bison 21 percent, ibex 9 percent, mammoth 8 percent, auroch 6 percent, deer 6 percent, reindeer 4 percent, lion, bear, and rhinoceros 2 percent each; and 12 percent assorted others. The selection of animals differs by cave; for example, at Chauvet, lions, mammoths, and rhinoceros are most numerous, whereas in the cave at Cosquer, horses and ibex are most numerous. By contrast, depictions of other animals such as hyenas, rabbits, rodents, snakes, birds, fish, and insects are rare to nonexistent. There is also no landscape included—the focus is solely on the animals.[23]

It is also noteworthy that the artists of this period attempted to depict the animals as realistically as possible. As noted by one art critic, "The artists seemed to aspire to represent a naturalistically convincing image of an animal, and their knowledge of its shape, stance, coat and expression indicates acute observation of the animals and their habits." For example, the painting of the spotted horses in the French cave at Pech Merle, dated to about 25,000 years ago, had been thought by some observers to be symbolic or fanciful. However, DNA studies of ancient horse bones recently confirmed that such dappled horses did indeed exist at that time, and the authors of the report concluded that such "prehistoric paintings were closely rooted in the real-life appearance of the animals depicted." Another example is drawings and paintings of walking horses. It is known that horses actually walk as follows: left hind, left fore, right hind, right fore. A study comparing depictions of horses by ancient cave artists and depictions of horses by recent artists from the past 200 years reported that the cave artists got the walking sequence correct more often than did the more recent artists. The author concluded that "cave painters understood—better than many artists of the modern age—the laws governing animal motion."[24]

The degree of artistic excellence achieved by many of the artists is impressive. In some cases, the natural contours of the cave wall were incorporated into the paintings. At Chauvet, for example, the horn of a rhinoceros follows the curve of the rock wall. In one panel, two rhinoceroses are squared off to fight; in another, a group of lions appears to be stalking prey; and a third panel, 30 feet wide, includes four horses, four bison, and three rhinoceroses. Twenty thousand years after Chauvet, the ceiling in the Altamira cave, dated to about 14,000 years ago, was covered with "21 magnificent painted bison outlined and shaded in black, red bodies engraved in the glistening, creamy limestone. They crouch, lie down, shake their manes, charge across the ceiling, heads turned, tails flying, drilled eyes dark as coal." As in the Chauvet Cave, the artists at Altamira took advantage of natural contours of the rock so that one bison, whose head is turned to look back, appears three-dimensional because its head is painted on an outcropping of rock. One senses that the artists had great respect, even reverence, for their subjects and that such paintings may represent an animal apotheosis. In this regard, there is also evidence that feasting took place at the entrance of the cave. Altamira Cave has been called the Sistine Chapel of Paleolithic art, and when Picasso visited it, he exclaimed: "We have invented nothing!"[25]

In contrast to the abundant animal figures, human figures in the cave art are relatively rare. Those that do exist are crudely drawn, often no more than stick figures. In some cases, the human figure is part of a hunting scene. For example, in Lascaux Cave, there is only one human figure among 915 animals. According to one analysis of the Lascaux drawing, "Apparently the man has wounded the bison, whose intestines are spilling out, and the bison has knocked over the man." Among the sculptures and figurines (portable art), human figures are more commonly found, especially the Venus figurines described earlier.[26]

Still another type of figure found among the cave art of this period is a composite human and animal figure, called an anthropozoomorph. One researcher claimed that there are more than 50 such figures "of men that are part animal and part human, or at least of men wearing animal disguises," but many of these figures are quite ambiguous. The best-known example, and probably the single most widely reproduced example of cave art, is a human-animal composite in Trois-Frères Cave in France; it is dated to approximately 15,000 years ago and was designated by one researcher as "the sorcerer of Les Trois Frères." Another well-known example is an ivory statue of a man with a lion's head, dated to about 40,000 years ago, found in a cave in southwest Germany, as mentioned earlier. The possible meaning of these figures will be discussed later.[27]

Although human figures are relatively rare in cave art, human handprints are very common, especially in the earlier painted caves. Chauvet Cave, with drawings dated to 36,000 years ago, has hundreds of handprints, the largest number being palm prints that are now faded and look like red dots. It also has complete handprints, made by covering the hand with pigment and then placing the hand against the wall, as well as "negatives" made by holding a hand against the wall and outlining it by blowing red ochre pigment over it. Gargas Cave in southwestern France, with drawings dated to 27,000 years ago, has over 200 handprints. The handprints in these caves are similar to those dating to later periods found in at least 30 other European caves and at rock art sites in South Africa, Indonesia, Australia, Papua New Guinea, Argentina, and the United States.[28]

In addition to animal figures and human handprints, the third category of commonly found figures in cave art is geometric figures. These are extremely numerous and found in almost all caves and rock art sites from this period. They vary from small dots and lines to circles and spirals to claviform (club-shaped) and tectiform (hut-shaped) drawings.

They are commonly included on panels that depict animals but also stand alone. In some cases, the geometric shapes are superimposed on the animals; straight lines so placed have been thought to represent spears or arrows. These geometric figures are often referred to as "signs" or "symbols," but it is unknown what they symbolize. They have been called "the most mysterious figures in cave art."[29]

Improved and novel forms of tools and weapons; memory devices; diversified and widespread self-ornamentation; intentional human burials with grave goods; musical instruments; painted caves; sculptures and figurines; decorated objects of all kinds—it was an outpouring of human creativity unlike anything seen in the six-million-year hominin history. As Randall White summarized it, "material forms of representation exploded onto the scene between 40,000 and 30,000 years ago in Europe." When these developments are placed on a timeline (see figure 5.1), many of them are seen to have occurred at approximately the same time in geographically disparate parts of the world. Some writers have referred to this period as a "human revolution."[30]

But was this really a "human revolution"? In 2000 two anthropologists, Sally McBrearty and Alison Brooks, published an influential article titled "The Revolution That Wasn't." They argued that many of the developments described as occurring about 40,000 years ago had, in fact, been seen 40,000 to 60,000 years earlier, including the use of bone tools, fishing, the use of body ornaments, human burials, and trade networks. Rather than being a revolution, they argued that it had been "an accretionary process, a gradual accumulation of modern behaviors" that had taken place primarily in Africa over 200,000 years.[31]

McBrearty and Brooks are correct that many of the developments seen beginning about 40,000 years ago had been seen thousands of years earlier, even if they were seen less commonly. I believe they are incorrect, however, in not viewing these changes as a human revolution. As noted in the previous chapter, it seems likely that a major cognitive

Years Ago	Tools and Weapons	Self-Ornamentation	Burials with Grave Goods	Arts
45,000		• pierced animal teeth (Bulgaria) • shell beads (Turkey, Lebanon)		• bone flutes: first-known musical instruments (Germany)
	• hooks and nets for deep sea fishing (Indonesia)	• ostrich eggshell beads (East Africa)		
40,000	• first-known lamps (France)	• shell and bone beads (Czech Republic)	• possible burials in Australia and South Africa	• sculpted ivory figurines of a lion, a man with a lion's head, and a woman (Germany) • geometric figure in cave in Spain and handprint in cave in Indonesia
	• bone tools and weapons become widespread			
35,000	• first-known needles for sewing clothing (Georgia)			• drawings of animals in Chauvet cave in France and in Leang Timpuseng cave in Indonesia
	• engraved memory devices (France) • ropes, baskets	• shell beads (Greece) • ivory and stone bead-making "factories" (France) with suggestions that beads were being used in extensive trade networks		
30,000	• spear throwers (France)			• rock art (Namibia) • rock art (Australia)
			• Sungir: first definite burial with grave goods (Russia)	• Venus of Dolni Vestonice: first-known ceramic object (Czech Republic)
			• Dolni Vestonice (Czech Republic)	• Venus of Willendorf (Austria)
			• Grimaldi cave (Italy)	• Gargas painted cave (France)

25,000		• pendants, bracelets, necklaces, headbands, headdresses, made using ivory, bone, seashells, antler, animal teeth, fish vertebrae, and various types of stone		• Pech Merle painted cave (France)
			• Mal'ta (Siberia)	
20,000	• lamps with engraved handles (France) • bow and arrow technology widespread • widespread use of spear points, shaft straighteners, chisels, wedges, awls, drills, spear throwers, many decorated with engravings		• burials with grave goods became increasingly common	
15,000				• Lascaux painted cave (France) • animal paintings on stones (Nambia) • incised designs on ostrich shells (South Africa) • Altamira painted cave (Spain) • rock shelter designs and paintings (India, Brazil)

FIGURE 5.1 Time Line: 45,000 to 13,000 years ago.

change—the acquisition of introspection—occurred approximately 100,000 years ago and that this cognitive change largely accounts for the new behaviors seen at that time. If that is true, is it possible that another major cognitive change took place approximately 40,000 years ago? If so, what was it?

MASTERING THE FUTURE: THE EVOLUTION OF AUTOBIOGRAPHICAL MEMORY

At about age four, children develop the first phases of what is known as autobiographical memory, sometimes also referred to as episodic memory. Prior to age four a child is said to live "in a comparatively foreshortened world of time. The present is what is outstanding for it. Its life goes neither far into the past nor far into the future." This was illustrated by experiments carried out by Daniel Povinelli and his colleagues in which they asked at what age young children "come to conceive of the self as possessing explicit temporal dimensions." They assessed time sense in children between ages two and five by placing a large sticker on their forehead, then showing the children videotapes of themselves with the sticker, with a delayed time interval between placing the sticker and showing the videotape. Almost none of the two- and three-year-old children reached up to remove the sticker, but most of the four year olds did. Povinelli et al. concluded that "the younger children possess a different understanding of the self than the older children. In particular, they may not readily appreciate that past events in which they participated (and hence of which they have memories) happened to them . . . although the events depicted may be recalled, they were not encoded as autobiographical memories and hence the children do not understand that they happened to them." As children get older, they are able "to knit together historical instances of him- or herself into a unique, unduplicated self." The result, in the words of psychologist and philosopher William James, is an "unbrokenness in the stream of selves," one able to project experiences from the past and present into the future.[32]

Studies of children also suggest that the development of basic cognitive skills, including an ability to compare a drawing or picture with what the person has seen in the past, may be necessary for an understanding of art. Such studies have demonstrated that children younger than two do not understand the nature of a picture; for example, they may try to pick up a ball that is in a picture. At age three, some children still believe that a picture of an ice cream cone will feel cold and a picture of a rose will smell sweet. Until age four, "many children think that turning a picture of a bowl of popcorn upside down will result in the depicted popcorn falling out of the bowl," according to University of Illinois psychologist Judy DeLoache, who has pioneered research on children's understanding of pictures. This suggests that the outpouring of art that occurred beginning about 40,000 years ago may have been dependent on the cognitive developments taking place at that time.[33]

Autobiographical memory is one of two types of long-term memory. Short-term, also called working, memory serves to "hold and handle information in the mind needed for the execution of cognitive tasks such as reasoning, comprehension, learning and carrying out sequences of action." Short-term memory is the memory used when you try to remember a new telephone number as you dial it. Long-term memory, by contrast, consists of memory "traces" that may be stored for decades. One type of long-term memory is called semantic memory. This is the long-term memory that stores facts, such as the capital of France. The second type of long- term memory is the autobiographical form. In contrast to semantic memory, autobiographical memory is a reliving of past events both sensually and emotionally. The difference has been described as follows: "It is our semantic memory that allows us to state the name and location of the high school we attended, [but] it is episodic [autobiographical] memory that allows us to re- experience the emotions and events during our first day at this school." Marcel Proust, in his *Remembrance of Things Past*, provided one of literature's best examples of autobiographical memory:

> One day in winter, as I came home, my mother, seeing that I was cold, offered me some tea, a thing I did not ordinarily take. . . . I raised to my lips a spoonful of the tea in which I had soaked a morsel of the cake. No sooner had the warm liquid, and the crumbs with it, touched my palate than a shudder ran through my whole body, and I stopped, intent upon the extraordinary changes that were taking place. An exquisite pleasure had invaded my senses, but individual, detached, with no suggestion of its origin. . . . And suddenly the memory returns. The taste was that of the little crumb of madeleine which on Sunday mornings at Combray . . . my aunt Léonie used to give me, dipping it first in her own cup of real or of lime- flower tea. . . . But when from a long-distant past nothing subsists . . . the smell and taste of things remain poised a long time, like souls, ready to remind us, waiting and hoping for their moment, amid the ruins of all the rest; and bear unfaltering, in the tiny and almost impalpable drop of their essence, the vast structure of recollection.[34]

Although researchers have largely focused on the past dimensions of autobiographical memory, there also is a future dimension. For example, your semantic memory will tell you the address of a four-star restaurant at which you have a reservation, but the future equivalent of your autobiographical memory will allow you to anticipate the visual and gustatory delights you hope to experience there. This has been called "pre-experiencing an event." Studies of the development of autobiographical memory in children have shown that the past and future dimensions develop simultaneously and are cognitively integrated. Together, they form the temporal self, enabling the person to use the past to master the future. This linking of the past with the future, according to Sir John Eccles, demonstrates "the extraordinary ability of humans to plan for the future while profiting from the memory of past experiences." Eccles added that "we live in a time paradigm of past-present-future. When humans are consciously aware of the time NOW, this experience contains not only the memory of past events, but also anticipated future events." It has even been claimed that "the primary role of episodic [autobiographical] memory . . . may be to provide information from the past for the simulation of the future."[35]

Several writers have noted the future as well as the past dimensions of autobiographical memory. The opening lines of T. S. Eliot's Four Quartets describe it succinctly:

> Time present and time past
> Are both present in time future,
> And time future contained in time past.

And in Lewis Carroll's *Through the Looking Glass*, the White Queen instructs Alice that "one's memory works both ways."

> "I'm sure mine only works one way," Alice remarked. "I can't remember things before they happen."
>
> "It's a poor sort of memory that only works backwards," the Queen remarked. "What sort of things do you remember best?" Alice ventured to ask.
>
> "Oh, things that happened the week after next," the Queen replied in a careless tone.[36]

Both semantic and autobiographical past memories may be lost in individuals who have Alzheimer's disease and, significantly, such individuals also lose their ability to envision the future. Occasional individuals have been described with other brain abnormalities who retain their semantic memory but lose their autobiographical memory. One such man "still had memories of brain facts," such as how to make a long-distance telephone call, but he "could not recall a single event from his own life." When asked about the future, he said his mind was just "blank"—"it's like being in a room with nothing there and having a guy tell you to go find a chair." Another man, who suffered brain damage following a heart attack, retained his semantic memory for past public events but "was unable to consciously bring to mind a single thing he had done or experienced before his heart attack." Thus, "he knew the name of the company where he had worked . . . but he could not recall a single occasion when he was at work or a single event that occurred there." Similarly, he cited global warming as a threat to the future but "had severe difficulty imagining what his experiences might be like in the future." The authors of this study concluded that autobiographical memory "enables a person to mentally travel back in time to relive previously experienced personal events," which in turn provides "a foundation for imagining what one's experiences might be like in the future."[37]

Do animals other than humans have an autobiographical memory? Many animals prepare for the future, storing food and migrating, but it is thought that they do so automatically, by instinct. Some researchers have claimed that chimpanzees have an ability to use the past to plan the future, since they have been known to save tools for possible future use. And then there is Santino, a chimpanzee at a Swedish zoo, who is said to collect stones in a pile so he can throw them at visitors when the zoo opens later in the morning. Other researchers have claimed that scrub jays have an autobiographical memory, since they not only store food but also anticipate when other birds are likely to steal their stored food. Most recently, some researchers have claimed that rats have an autobiographical memory based on the activation of their hippocampus when they are running mazes. The debate whether such behaviors represent true autobiographical memory continues, with the majority of researchers calling the evidence equivocal.[38]

The acquisition of autobiographical memory by modern *Homo sapiens* would have provided them with a significant evolutionary advantage over Neandertals and the other remaining species of Archaic *Homo sapiens*, who apparently did not possess this cognitive skill. It allowed humans to flexibly consider a variety of past events in planning future behaviors. To illustrate, consider the difference between hunters equipped with only semantic memory 75,000 years ago and hunters equipped with both semantic and autobiographical memory 25,000 years ago. A hunter 75,000 years ago might have planned as follows: "I remember that the reindeer came down the valley and crossed the river when the sun went down over the hill. I killed two of them and will hunt again next year."

In contrast, a hunter 25,000 years ago might have planned as follows:

> I remember that the reindeer came down the valley and crossed the river at the time when the sun went down next to the large tree on the hill, because that was the same time my sister died giving birth. We only killed twelve reindeer because my brother-in-law's clan, with which we were hunting, brought along young boys who made too much noise and couldn't follow orders. So next fall, we will not hunt with them but rather with the clan of my mother's sister. And we will station women downstream around the bend of the river to drag the dying reindeer out of the water so the men don't have to take the time to do that but can keep on killing the reindeer. My brother-in-law may be angry with me, but I will give him my fox-teeth pendant, which he has admired in the past, so then we will remain on good terms. If we plan the hunt carefully and give everyone assigned tasks, we should be able to kill thirty or more reindeer. This will give us good food to store for winter.

This hypothetical scenario illustrating the advantages of an autobiographical memory is consistent with the fact that hunting at this time is said to have "shifted from hunting individual and small groups of animals to slaughtering mass herds of reindeer and red deer . . . likely to have been attacked at critical points on their annual migration routes when the animals were constrained in narrow valleys, or when crossing rivers." Modern humans were probably keeping precise records telling them when the animal migration was likely to occur, and they could then envision various scenarios predicting where the animals would be most vulnerable. Thus, during the spring salmon run or annual deer migrations, modern *Homo sapiens* could use their autobiographical memory to maximize their acquisition of food.[39]

There are additional suggestions in the later years of this era, between approximately 18,000 and 11,000 years ago, that such mass killings of animals were carried out by large groups of people in a cooperative manner. Instead of each small group of hunter-gatherers going its own way, they increasingly joined together at preappointed times to hunt cooperatively. As South African archeologist David Lewis-Williams described it: "We also need to note the presence of large Upper Paleolithic settlements. . . . These settlements were probably aggregation sites. Communities split up into small bands during some seasons of the year and then united at recognized aggregation sites at others." Some of these aggregation sites have been identified in France and Spain, such as the area around the cave at Altamira, and include structures suggesting that large numbers of people lived in what appear to have been permanent shelters.[40]

Such cooperative hunting would also have been facilitated by the combination of autobiographical memory with language. As Australian psychologist Thomas Suddendorf and colleagues noted: "The evolution of language itself is intimately connected with the evolution of mental time travel. . . . Language allows personal episodes and plans to be shared, enhancing the ability to plan and construct viable futures." Suddendorf even claimed that "mental time travel was a prime mover in human evolution."[41]

THE EMERGENCE OF RELIGIOUS THOUGHT 1: THE MEANING OF DEATH

Using an evolutionary perspective, the emergence of religious thought was described by British anthropologist Edward B. Tylor in his book *Primitive Culture*, published in 1871. Tylor had been strongly influenced by Charles Darwin and his publication *On the Origin of Species*, published in 1859. As Darwin theorized that modern *Homo sapiens* had evolved from earlier hominins and primates, so Tylor theorized that "higher" cultures had evolved from "lower" or "primitive" cultures such as those Tylor had studied in Mexico. Tylor and Darwin corresponded, and Tylor cited Darwin's cultural findings in his book. Tylor believed that "primitive" people had initially developed religious ideas based on their understanding of death and dreams. Such an understanding would have been made possible by the acquisition of an autobiographical memory.[42]

Prior to about 40,000 years ago, hominins had been observing other hominins die for more than six million years. They were intimately acquainted with death as something that happened to others. They observed people die within their living group—children from disease, women from childbirth, men from hunting accidents, and older adults from starvation. They also occasionally encountered deceased hominins as they foraged for food or followed herds of deer. Unlike today, when the biological realities of death are relegated to the offices of medical examiners and morticians, early hominins saw corpses in all stages of decomposition, since even the occasional burial of bodies was apparently not practiced until the last 100,000 years.

What did these early hominins observe? Within hours of death, a person's skin becomes blotchy in areas where the blood has settled and ashen elsewhere. Rigor mortis stiffens the muscles for a couple of days, by which time decomposition has begun. The first organ to go is the brain, which breaks down to amino acids and lipids and becomes a viscous gray liquid that may seep out of the person's ears, nose, or mouth.

Decay of the rest of the body usually begins by the third day and comes from both within and without. Within the intestines, millions of bacteria, which were previously held in check by the body's immune system, digest the intestine and other organs. In so doing, they produce gas, which bloats the body, especially the stomach, male genitals, lips, and tongue, which may cause it to stick out of the mouth. From outside the body, maggots collect around the eyes, mouth, and genitalia and begin to digest subcutaneous fat.

By the end of one week, increased bloating causes the internal organs to rupture. The skin becomes greenish, and in some areas peels off. Maggots, which by then are visible over most of the body, may be joined by beetles, which favor muscle tissue. By the end of two weeks, corpses are said to "basically dissolve; they collapse and sink in upon themselves and eventually seep out onto the ground." The smell of decaying flesh, noticeable at some distance, "is dense and cloying . . . halfway between rotting fruit and rotting meat"; the smell is said to be "poignant and memorable." Within two to four weeks, depending on how warm the temperature is, a corpse is reduced to a skeleton. The bones will also decompose, although this may take several years to complete. During this period, the bones and skull sit there, wretched reminders to the living.

British physician and philosopher Raymond Tallis described this period:

> Your skull, meanwhile, is as hospitable to these insects as it is to the thoughts you are presently having about them. That is what its dumb hardness, that you feel now, says: your head is not on anyone's side, least of all yours. It is as indifferent to your sorrows, your fears, your joys as it is to the song of the birds that might one day find it a ready-built shelter, as hospitable to the snake that slithers through your orbital fissure as to the light from which you constructed the image of your beloved. And not one of the creatures that grow, hop or gnaw their way through your rotting head will be the slightest bit curious about your thoughts, however privileged, original or salacious.[43]

All of this assumes that the body has not been disturbed by scavengers, which must have been an exception in the past. Scavengers, such as hyenas, selectively eat the arms and legs, which contain large muscles and long bones, the marrow of which is especially sought after.

Since these early hominins had observed decomposing bodies, they would have been acutely aware of the fact of death. And when people with whom they were closely associated died, they would have felt sadness and grieved, just as many animals do. The feeling of sadness and empathy may also explain why some Neandertals buried their dead, as a sign of caring or to protect the bodies from predators. Death would have been a fact, like the sun dying at the end of the day and warm weather dying at the end of the summer. Death was something that happened to other people; to understand that it is also going to happen to you, you need to be able to fully project yourself into the future, both theoretically and emotionally, using your accumulated experiences from the past. In short, one needs to have acquired an autobiographical memory.

As modern *Homo sapiens* slowly developed an autobiographical memory, an awareness of their own death began to take hold. Since they were able to introspectively think about their own thinking, entirely new ideas were born—infinity, eternity, the meaning of life. Once burdened with such ideas, it was no longer possible for one human to pass the decaying corpse of a fellow human without being assailed by unbidden questions. What happened to this man whom I knew? Where did he go? Will this also happen to me? Where will I go? Will I simply rot and dissolve into the ground, like this man? As Hamlet said of Caesar: "Imperious Caesar, dead and turn'd to clay, / Might stop a hole to keep the wind away." *Homo sapiens* would never again be free to not ask such questions. In the words of Theodosius Dobzhansky: "A being who knows that he will die arose from ancestors who did not know."[44]

Therefore, at the same time that the acquisition of autobiographical memory had conferred significant evolutionary advantages, it had also carried with it a massive millstone. Since modern *Homo sapiens* could both introspectively think about themselves and also project themselves into the future, they became fully aware, for the first time in history, that they were going to die. Thus, modern *Homo sapiens* became the first hominin to fully understand the implications and meaning of death. According to British archeologist Mike Parker Pearson, this awareness is "a fundamental defining characteristic of what it is to be human, at the very core of our being and self-consciousness." As theologian Paul Tillich phrased it, "The anxiety of death is the most basic, most universal and inescapable." Fear of death was the theme of the world's first recorded story, *The Epic of Gilgamesh*, which will be discussed in chapter 7, and it has continued to permeate literature to the present. As French poet Charles Baudelaire described it:

> Nothing can withstand the Irreparable—
> its termites undermine
> our soul, pathetic citadel, until

the ruined tower falls
Nothing can withstand the Irreparable!

According to Vladimir Nabokov: "The cradle rocks above an abyss, and common sense tells us that our existence is but a brief crack of light between two eternities of darkness." And T. S. Eliot captured it in a single line: "I will show you fear in a handful of dust."[45]

Edward Tylor theorized that "primitive" people, faced with an understanding of death, would have reasoned that something had been lost in the transition from being alive to being dead. That something, said Tylor, was a soul or spirit. And "if the concept of a soul explains the movements, activities, and changes in a human person, why should it not also be applied more widely to explain the rest of the natural world?" Tylor believed that a belief in souls or spirits was the essence of religious thought and called his theory animism, from *anima*, the Latin word for "spirit." Tylor even defined religion simply as a "belief in spiritual beings."[46]

There are suggestions in child development that a mature understanding of death is a relatively late acquisition in human evolution. Most children younger than six have no understanding of death. In his evocative memoir of childhood, Vladimir Nabokov reflected this: "A sense of security, of well-being, of summer warmth pervades my memory. That robust reality makes a ghost of the present. The mirror brims with brightness; a bumblebee has entered the room and bumps against the ceiling. Everything is as it should be, nothing will ever change, nobody will ever die." Young children believe that death is reversible, like going to sleep, and that people who die may return. In a study of 378 young children, many were said to believe that dead people continue to eat, drink, and experience thoughts and emotions. Between the ages of six and nine, a child's concept of death becomes more personified and frightening; death is described as being like a skeleton, but it is still neither permanent nor personal.[47]

A mature understanding of death does not begin to develop until the age of nine or later and involves four concepts: that death is universal; that it is irreversible; that all bodily functions cease; and that it has physical causes. A 10-year-old girl, for example, described death as "the passing of the body . . . like the withering of flowers." Even some adolescents, however, do not appear to fully understand death, as measured by their risk-taking behaviors. Thus, a mature understanding of death appears to be one of the last milestones in the cognitive development and evolution of the human brain.[48]

It is also of interest that no animal except modern *Homo sapiens* appears to fully understand death, suggesting that such understanding requires the development of autobiographical memory. Some animals may display evidence of bereavement, such as a dog may do when its master dies. Elephants have also been described showing signs of what appears to be bereavement, such as running their trunks over the bodies of dead family members and even throwing dirt on them. But grieving another's death is not the same as understanding that you too will die.[49]

Even among chimpanzees, the primates most closely related to humans, there is no indication that they understand death. Jane Goodall recorded 66 chimpanzee deaths in Tanzania and saw the bodies of 24 of them. In most cases, the dead animal was simply ignored and left to decay. In one case, when an adult male died after falling from a tree and breaking its neck, "group members showed intense excitement and anxiety, displayed around the body, and threw stones at it." Three other chimpanzee deaths occurred when adults killed infants and ate them. Cannibalism of dead members of their species has also been observed among gorillas, baboons, and other primates. Since an understanding of death appears to be unique to humans, it has even been suggested that "the knowledge of death is a much more decisive break between human modality and animal existence than tool-production, brain, or language."[50]

Many observers over the years have viewed an awareness of death as an impetus to religious thought. In ancient Rome, Gaius Petronius said: "It is fear [that] first created gods in the world." More recently, English philosopher Thomas Hobbes noted in *Leviathan* that religion is found "in man only" and reasoned that "the seed of Religion" must consist of "some peculiar quality . . . not to be found in other Living creatures." That "peculiar quality," said Hobbes, is the ability of man "which looks too far before him, in the care of future time, hath his heart all the day long, gnawed on by feare of death . . . the Gods were at first created by humane feare." Thus, the modern hominin who emerged beginning about 40,000 years ago was significantly different from all hominins that had previously lived. He was, in the words of Erich Fromm, "an anomaly, the freak of the universe . . . part of nature, subject to her physical laws and unable to change them, yet he transcends nature." An awareness of death was an inevitable by-product of our introspective and temporal selves, which, in themselves, conveyed enormous evolutionary advantages. To be fully human and to be aware of death are one and the same thing. In the words of William Butler Yeats: "He knows death to the bone— / Man has created death."[51]

To say that an awareness of death was the original impetus to religious ideas 40,000 years ago, however, is not to say that a fear of death dominates the thinking of modern *Homo sapiens*. This later position has been proposed by some social psychologists, based on Ernest Becker's contention in 1972 that a fear of death "is a mainspring of human activity—activity designed largely to avoid the fatality of death, to overcome it by denying in some way that it is the final destiny for men." As phrased more recently: "All human activities are framed by death anxiety and colored by our collective and individual efforts to resolve the inescapable and intractable existential given."[52]

The social psychologists who have developed this theory call it Terror Management Theory and suggest that we use self-esteem and our cultural worldview to buffer our anxiety about death. Supporters of this theory have argued that it can be scientifically tested by reminding people of their own death (called "mortality salience") and then measuring the effect of this reminder on their thinking. A summary of 277 such experiments claimed that evidence to support Terror Management

Theory "is robust and produces moderate to large effects across a wide variety of MS [mortality salience] manipulations."[53]

Other researchers have criticized Terror Management Theory. They contend that a person's self-esteem and cultural worldview are shaped by many factors other than death anxiety. They also criticize the "mortality salience" experiments for the methods used to assess death anxiety. Most important, it is questionable whether contemporary Terror Management Theory, based on a cultural worldview in which virtually everyone as a group accepts the existence of an afterlife, has much relevance for *Homo sapiens* who lived 40,000 years ago and were becoming aware of their own death for the first time.

THE EMERGENCE OF RELIGIOUS THOUGHT 2: THE MEANING OF DREAMS

The belief that all humans have a soul or spirit, and that this soul leaves the body at the time of death, was merely the first part of Edward Tylor's theory regarding the origin of religious thought. The second part of his theory was "the belief in the soul's continued existence in a Life after Death." Tylor contended that "primitive people arrived at this conclusion based upon their experience with dreams."[54]

What do we know about dreams? We know that they are associated with rapid-eye-movement (REM) sleep, and that all mammals have periods of REM sleep. Dogs, cats, monkeys, and elephants are said to dream, and perhaps all mammals do. The purpose of REM sleep and dreams is still unknown; theories have included functions associated with memory storage, problem solving, and threat simulation. Some researchers have theorized that REM sleep and dreams are evolutionary epiphenomena that perhaps played some useful role in our distant past.

However, if hominins had been dreaming for several million years, why did dreams become more important about 40,000 years ago? The reason is that hominins could not assign meaning to their dreams until they had cognitively matured. Specifically, they needed to have acquired an awareness of self, awareness of others, introspection, and an ability to place the experience of their dream within the context of their past experiences and future hopes.

Anthropologist A. Irving Hallowell referred to the necessity of cognitive maturation in interpreting the dreams of Ojibwa Indians: "Dreaming may have occurred in the early hominids, but, without the psychological potentialities fully released only with the expansion of the hominid brain, it would not have been possible for the content of dreams or the products of imaginative processes to have been communicated to others."[55] Edward Tylor theorized that the "primitive" people's experience with dreams led them to the idea that the soul or spirit that leaves the body at death continues to live in some kind of spirit world or land of the dead. He cited two kinds of dreams as having been especially important in fostering the idea of an afterlife. The first was dreams in which "human souls come from without to visit the sleeper who sees them as dreams." Tylor cited as examples the Zulu of South Africa, who "may be visited in a dream by the shade of an ancestor," and people in Guinea in West Africa, whose "dreams are construed into visits from the spirits of their deceased friends." The other kind of dream described by Tylor was that in which a person's soul leaves their body during sleep and travels to other places, including the land of the dead. Thus Tylor cited the Maoris of New Zealand, whose dreaming souls could "leave the body and return, even traveling to the region of the dead to hold converse with its friends." Given such evidence from their dreams, Tylor argued that it was "rational enough from the savage point of view" to conclude that there is "an independent existence of the personal soul after the death of the body, in a Future Life."[56]

The importance of dreams in shaping religious ideas has been widely noted. For example, Boston University neurologist and researcher Patrick McNamara cited "the importance of the dream as a primary source for religious ideas and practices of traditional peoples."

> Both ancestral and nonancestral supernatural agents appear in dreams and are reverenced in daily life. The spirit beings that appear

> in dreams can be either positively or negatively disposed toward the dreamers—that is, both evil and good supernatural beings appear in dreams. . . . Dream characters, therefore, have a prima facie case to be considered as the cognitive source for supernatural beings. People in traditional societies treat them as such, and it is likely that ancestral populations also treated them as such.

Therefore, "there can be little doubt that dream experiences have been thoroughly intertwined with the religious beliefs, practices, and experiences of people all over the world, throughout history."[57]

A review of ideas about dreams in contemporary anthropological accounts supports McNamara's conclusion. Among 295 cultures described in the Human Relations Area Files (HRAF) at Yale University, 71 of them are economically categorized as being primarily or largely dependent on hunting, gathering, and fishing for their livelihood. In the available ethnographic accounts of these cultures, the importance of dreams is mentioned in all but two cultures. Sometimes the dreams are mentioned as predicting the future; at other times the dreams include visitations by deceased relatives or visits by the dreamer to the land of the dead. For example, among the Nootkan Indians of Western Canada: "People frequently see the dead in dreams and this is regarded as good evidence for the nature of the life of the dead." And among the Mataco Indians of Bolivia: "Very often in dreams one sees dead relatives. The soul has gone to the underworld and paid them a visit." Appendix B lists 25 accounts of dreams taken from hunter-gatherer cultures from all over the world as described in the HRAF files.[58]

The nature of the spirit world or land of the dead, as envisioned by different cultures, varies widely. For example, Native American Pawnee believed that "the soul of the deceased ascended to heaven to become a star." The souls of the Yakut in Siberia were said to "travel skyward to a lush greenery-filled heaven." The spirits of dead Yanomama in Brazil went to the sky, which "resembles earth except the hunting is better, the food tastier, and the spirits of the people are young and beautiful." The souls of Aboriginal Australians were said to go to "a beautiful country above the clouds abounding with kangaroos and other game." In a few cultures, the afterlife was located underground. For example, in Samoa, the entrance to the afterworld was through an active volcano, and among the Chukchee in Siberia, the dead lived underground, where "the reindeer herds are numerous."[59]

Thus, about 40,000 years ago, an idea slowly took hold that human spirits continue to live after the human body dies. The development of this idea occurred over thousands of years as brains evolved an introspective and temporal self and modern *Homo sapiens* became increasingly uneasy at the prospect of their own death. This was not merely a semantic denial of death, as when we say "passing on" rather than "dying." This was a basic conceptual denial of the fact that death is the termination of our existence. In place of an embrace by maggots and betrothal to dust, we discovered an afterlife in which we continue to exist in other forms, as a spirit or soul in an afterworld, or are reincarnated in another body or form. Humans became, for the first time, immortal.

THE HUMAN REVOLUTION REVISITED

The evolution of autobiographical memory, allowing modern *Homo sapiens* to more skillfully utilize the past to plan the future, could largely explain the human revolution that began approximately 40,000 years ago. Tools and weapons would have been rapidly improved as humans incorporated their experiences from the past in planning for their needs in the future. The widespread use of memory devices at this time reflects an increased interest in keeping records for past events, such as the number of reindeer killed last fall, and predicting future events, such as the next full moon.

In view of the development of autobiographical memory, how should we interpret the outpouring of visual art at this time? In interpreting the art, however, it is important to keep in mind that our theorizing is based solely on the art that has survived and been discovered. As Paleolithic art expert Jean Clottes noted, "It is certain that we know but a fraction of the caves which were painted and/or engraved."[60]

The main theme of the art of this period, as noted, is animals, especially those being hunted. Such animals were central to the survival of these people; for example, wild horses and reindeer were "the foundation of the humans' diet." Thus, the most parsimonious explanation is that the artists were depicting things they had seen in the past or hoped to see in the future. Such an explanation is supported by the fact that "15 percent of these animal paintings depict animals that have been wounded by spears or arrows," thus suggesting they are hunting scenes. Such paintings may also have been used to teach children about animals and how to hunt, and children's footprints have been found in several of the caves.[61]

What about the handprints in the caves? Perhaps they are the equivalent of Paleolithic graffiti, the universal way of saying, "I was here." Such graffiti are a record of where the person was in the past as well as a message for future observers. That would be consistent with the newfound ability of modern *Homo sapiens* to project themselves into the future. Niaux Cave in southwestern France has paintings dated to about 13,000 years ago, but it also has graffiti left by visitors to the cave in recent centuries, including an individual named "Ruben de la Vialle," who left his name in the cave in 1660. Is this really different in intent from the much earlier handprints?[62]

It is also possible that some of the art reflects the newly developing religious ideas, especially the belief in spirits. Given the abundance of animals, some of the art may represent the spirits of animals. A belief in animal spirits is widespread in the world. In some cultures, it is believed that the animal is the ancestor of humans; such an animal is called a totem and is found most strikingly among Australian Aborigines and Northwest Coast Indians. Adherents of a totemic interpretation of the painted caves point to the predominance of specific animals in specific caves and to the half-human, half-animal figures. They also point to a collection of bear skulls in Chauvet Cave, one of which was prominently placed on a rock. As one scholar summarized the scene:

"Somehow, some way, this room was a shrine to cave bears, which were honored in unknowable rituals."[63]

If animal or other spirits were present in the painted caves, they may have also been used to explain the unknown. Providing explanations for events that are not understood is a function of almost every religious system that has been described. Thus, the animal spirits may have been invoked to explain why the reindeer were late that year in arriving at the river crossing, or why a young man suddenly became sick the day after a bear crossed the trail in front of him.

More elaborate religious explanations of the painted caves have also been proposed. French prehistorian Jean Clottes is regarded as a world authority on prehistoric art, and in 1998 Clottes and South African archeologist David Lewis-Williams published *The Shamans of Prehistory*. Then, in 2003 Canadian archeologist Brian Hayden published *Shamans, Sorcerers, and Saints*, expanding on this theme. *Shaman* was originally a precise term denoting the indigenous healers among the Tungusian tribes in Siberia who healed by undergoing a trance state. Subsequently, it has been used much more broadly to refer to seers who can foretell the future or control the weather, sorcerers who can cast spells, and priest-like individuals who act as intermediaries between this world and the afterworld. The use of *shaman* in discussions of the painted caves appears to focus mostly on their priest-like function.[64]

The shamanistic interpretation of cave art postulates that priest-like shamans existed at the time when the caves were painted and that much of the cave art was a product of their trance states. The caves themselves are said to be passages to the underworld of the dead so that people who walked through the caves "were completely surrounded by the underworld." The handprints on the cave walls are interpreted as attempts by people to contact the underworld. The geometric figures are said to represent visual hallucinations seen by the shamans while in a trance. Specific parts of the cave are said to have been designated for various spiritual functions, including a meeting place for secret societies. The half-human, half-animal figures are said to represent the shamans. Abbé Henri Breuil, a French Catholic priest who spent his life studying cave art, originally designated the most famous one as the "sorcerer of Trois Frères" but later called it the "god of Trois Frères."[65]

But were gods present in the painted caves? As discussed in the preface, the term *gods* has sometimes been used very broadly to include any kind of supernatural beings, including animal spirits. If such a broad definition is used, then probably gods were present in the painted caves. However, if *gods* is defined more narrowly, as the term is more commonly used, to indicate male or female divine beings who are immortal and who have some special powers over human lives and nature, then it seems less likely that gods were present in the painted caves.

What about religion—was it present in the painted caves? As noted in the preface, the answer to this depends on which of the many definitions of religion you select. Edward Tylor defined religion broadly as "the belief in spirit beings," thus qualifying as religion any activities in the painted caves that involved spirits, animal or otherwise. Similarly, using the broad definition proposed by French sociologist Émile Durkheim, the worship of totems would qualify as a religion, since, according to Durkheim, "a religion is a unified system of beliefs and practices relative to sacred things." In fact, Durkheim regarded the worship of totems as the "simplest and original form" of religion, and he studied and wrote about totem worship among Australian Aborigines. If, on the other hand, a religion is defined more narrowly, as William James suggested, as "the feelings, acts, and experiences of individual men . . . as they apprehend themselves to stand in relation to whatever they may consider the divine," with divine said to be "godlike," then it seems less likely that the spiritual activities in the painted caves would have qualified as a religion.[66]

Given this broad range of interpretations of the meaning of the cave art, what is likely to be the correct one? In fact, it seems unlikely that there ever was a single correct interpretation. The art covers more than 20,000 years, a period 10 times longer than the period from the birth of Jesus to the present. The creation of the art in any given cave often covered several centuries— Chauvet Cave approximately 8,000 years, Cosquer Cave 6,000 years, and even Lascaux Cave, in which most of the art is similar, up to 1,000 years. Given such time spans, attempts to interpret the precise placement of specific art within a cave, as if it had been arranged all at once by an interior decorator, seem futile.[67]

At a minimum, the cave art would appear to depict things the artists had seen in the past or hoped to see in the future. The animals, especially those with arrows or spears in them, may be attempts to magically ensure the success of the hunt. Such attempts at "hunting magic" have been described by anthropologists among contemporary groups of hunter-gatherers, and this interpretation of the cave art was popular in the last century.[68]

It is also possible that the animals depicted were believed to be totems, ancestral spirits of humans. This is especially likely to have been the case in the later years of this period. Some of the animals depicted in the last painted caves even overlap in time the animals depicted at Göblecki Tepe beginning approximately 11,500 years ago, by which time the evidence for ancestor worship by humans is stronger, as will be discussed in the next chapter.

However, the evidence for a shamanistic or more elaborate religious activity in the painted caves is questionable. It seems premature to call a half-human, half-animal figure a "god" or to postulate the existence of secret societies, presumably based on shared religious beliefs. Such things are possible, but the existing evidence does not appear to support them. Such excessive religious speculation of cave art has also been criticized by archeologist Dale Guthrie as denigrating to the people of that period: "This magico-religious paradigm . . . has presented early peoples in a distorted light as superstitious dolts totally preoccupied with mystical concerns. Yet the evidence from Paleolithic art tells a quite different story; it portrays people in close touch with the details of a complex earth. Religious images probably do occur, but they are part of a larger mosaic of experience."[69]

In view of the development of autobiographical memory, how should we regard the placement of valuable goods in burials, a practice that was becoming widespread at this time? Goods may be placed in graves for many reasons, as Edward Tylor noted more than a century ago. Such reasons include burying the deceased person's favorite personal items; burying an item as a sign of affection for the deceased; and burying the deceased person's personal items so that their spirit will not return to the house looking for them.[70]

The most common reason for burying personal items in a grave, however, is so that those items will be available for use by the deceased in the afterlife. As Ian Tattersall observed, "Burial of the dead with grave goods . . . indicates a belief in an afterlife: the goods are there because they will be useful to the deceased in the future." Similarly, Steven Mithen has argued that "it is difficult to believe that such investment would have been made in burial ritual, as at Sungir, had there been no concept of death as a transition to a non-physical form." This interpretation of grave goods is also consistent with the practice by some groups of killing people so that they can serve the deceased in the afterlife. In his *Primitive Culture*, Tylor included many examples of this practice prior to the arrival of missionaries. For example, "the Caribs . . . sacrificed slaves on a chief's grave to serve him in the new life, and for the same purpose buried dogs with him, and also weapons." Grave goods are thus one of the most dramatic examples of the effects of autobiographical memory and its associated belief that humans continue to live in another world after their body dies.[71]

THE BRAIN OF MODERN *HOMO SAPIENS*

As modern *Homo sapiens* was gradually acquiring a fully mature autobiographical memory approximately 40,000 years ago, what was happening to their brain? Newly expanding brain areas were acquiring new functions, older brain areas were being reprogrammed, and the white matter connections between them were being improved. The temporal self and the brain of modern *Homo sapiens* were evolving together.

Substantial research has been carried out to identify the brain areas that are associated with autobiographical memory. To assess this, individuals are subjected to brain imaging while being asked to focus on specific kinds of memories. A review of 19 such studies identified multiple brain areas that were highly activated, as seen in figure 5.2. Several of the identified areas were identical to areas activated by thinking about others (theory of mind), described in chapter 3. These include the anterior cingulate (BA 24, 32), part of the inferior parietal lobule (BA 39), and the adjacent posterior superior temporal area (BA 22). One study, for example, reported that the "inferior parietal cortex was particularly active during retrieval of self-referential information."[72]

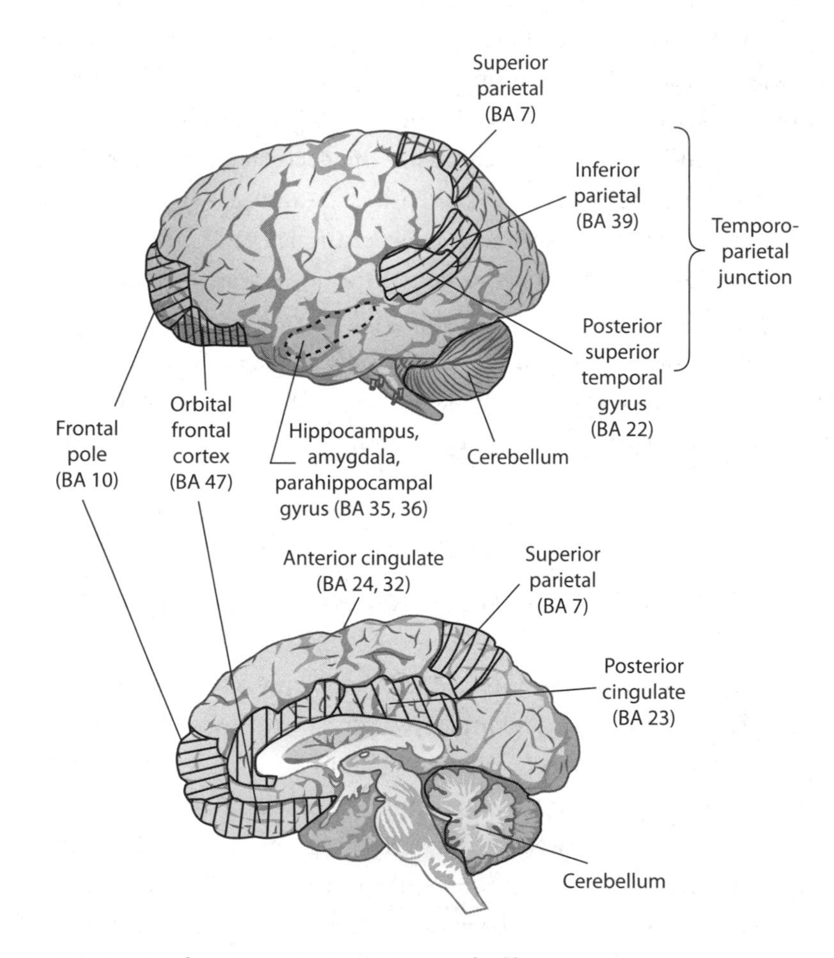

FIGURE 5.2 Modern *Homo Sapiens*: a temporal self.

The prefrontal cortex is also activated by autobiographical memory tasks, especially the frontal pole (BA 10) and orbital frontal cortex (BA 47). Studies of the brains of individuals with frontotemporal dementia, in which almost all autobiographical memory is lost, have reported that the orbital frontal cortex is severely damaged. A study of young adults in which subjects were asked to think about the details of their first day at school, first kiss, and so on reported that the "medial prefrontal cortex (MPFC) was specifically engaged in the retrieval of recent AMs [autobiographical memories]." A review of frontal lobe function similarly reported that "the frontal lobes, in particular the frontal poles, are involved in uniquely human capacities, including self-awareness and mental time travel." This is also consistent with the fact that individuals who have sustained damage to their prefrontal cortex have impairments in thinking about the future equally as often as they have problems in thinking about the past. As summarized by one research group: "Frontal-lobe-damaged patients respond only to concrete, present situations, having no thought or plans for the future. . . . Many of the patients described appear to live entirely in the present. Implications of the past and the relation to decisions in the future appear to be significantly hindered."[73]

Some of the other brain areas that are activated by autobiographical memory tasks overlap only minimally the areas that are activated by the cognitive tasks previously discussed. Foremost among the other areas associated with autobiographical memory are the hippocampus (which has no BA number) and parahippocampal gyrus (BA 35, 36). Since these are evolutionarily among the oldest brain areas, it is somewhat surprising to find them involved in autobiographical memory, one of the evolutionarily newest human cognitive functions. Since the hippocampus is critical for memory storage, this is an example of an evolutionarily newer brain function co-opting an older brain area for its own purposes. Similarly, the evolutionarily older amygdala is involved in autobiographical memory because such memories are often emotion-laden, and emotion is a function of the amygdala. The hippocampus and amygdala are thus both crucial components of autobiographical memory.[74]

The importance of the hippocampus, amygdala, and parahippocampal gyrus for the development of autobiographical memory is also suggested by the development of the white matter connecting tracts. The cingulum is a major tract that connects the hippocampus, amygdala, and parahippocampal gyrus to the frontal lobe and parietal lobe structures involved in autobiographical memory. The uncinate fasciculus is also important in connecting many of the autobiographical memory brain areas. It is thus of interest that the cingulum and uncinate fasciculus are regarded as being the two most recently developed white matter connecting tracts in humans, consistent with the recent development of our autobiographical memory.[75]

Additional brain areas activated by autobiographical memory tasks include the posterior cingulate (BA 23, 31), the adjacent superior parietal area (precuneus, BA 7), and the cerebellum. The posterior cingulate, as noted in chapter 4, plays an important role in introspection. The precuneus, as noted in chapter 1, performs a variety of cognitive, sensory, and visual functions. The finding of the cerebellar involvement in autobiographical memory is somewhat surprising, since it is a very old brain area thought to be primarily associated with motor functions, such as the coordination of movements. However, in modern *Homo sapiens* the cerebellum has undergone "a rapid expansion . . . in modern brains" and is now three times larger in modern humans than in chimpanzees of comparable size. A study in which subjects were asked to recall specific past memories produced extensive cerebellar activation, and it was concluded that the cerebellum is part of a network "that initiates and monitors the conscious retrieval of episodic [autobiographical] memory." The fact that the cerebellum plays such a role is another example of an older brain area being co-opted for a more recently evolved brain function.[76] In fact, the more the cerebellum is studied, the more functions it is found to have.

Finally, are there differences between the brain areas activated by autobiographical memories of the past and areas activated by imagining future events? Several studies have addressed this question and reported remarkably consistent results: the brain areas activated by the two tasks are almost identical. One study noted that "this striking neural overlap is consistent with findings that amnesic patients exhibit deficits in both past and future thinking, and confirms that the episodic [autobiographical memory] system contributes importantly to imaging the future." They thus concluded that "neuroimaging studies from our laboratory and others reveal striking commonalities in the brain networks that are activated when people remember past episodes and imagine future ones."[77]

In summary, by about 40,000 years ago, hominins had completed five important stages of cognitive evolution. As *Homo habilis*, about two million years ago, they began to become significantly more intelligent, and this trajectory toward becoming progressively smarter continued. As *Homo erectus*, about 1.8 million years ago, they developed self-awareness. As the Neandertal species of Archaic *Homo sapiens*, about 200,000 years ago, they acquired an awareness of what others were thinking, called a theory of mind. As early *Homo sapiens*, about 100,000 years ago, they acquired an introspective ability to think about themselves thinking about themselves. Finally, as modern *Homo sapiens*, about 40,000 years ago, they developed an autobiographical memory, the ability to project themselves backward and forward in time, using their experiences from the past to plan the future. Each stage in this cognitive evolution was accompanied by anatomical changes in their brain, changes that, at least in broad outline, can now be identified.

Thus cognitively equipped, modern *Homo sapiens* was ready to domesticate plants and animals and to create states and civilizations. It would be an extraordinary series of developments. Accompanying these developments, however, would always be the eternal questions lingering in the shadows. "Where did I come from?" "Why am I here?" "What will happen to me after I die?" Modern *Homo sapiens* would find answers to such questions in our gods and religions.

2

THE EMERGENCE OF THE GODS

6

ANCESTORS AND AGRICULTURE

A Spiritual Self

Perhaps the whole root of our trouble, the human trouble,
is that we will sacrifice all the beauty of our lives,
will imprison ourselves in totems, taboos, crosses,
blood sacrifices, steeples, mosques, races, armies,
flags, nations, in order to deny the fact of death,
which is the only fact we have.

—JAMES BALDWIN, *LETTER FROM A REGION OF MY MIND*, 1962

The millennium from 12,000 to 11,000 years ago is generally regarded as a fault line in human history. It is the traditional dividing line between what is called the Paleolithic period and the emerging Neolithic or agricultural revolution, during which *Homo sapiens* began the transition from being hunters and gatherers to being settled farmers. During this period, the domestication of plants and animals began, an event said to be "the single most important feature of the human domination of our planet."[1]

A changing climate played an important role in this transition. The last ice age began about 25,000 years ago, peaked about 18,000 years ago, and then slowly began to warm.

Glaciers, which had covered much of the northern hemisphere, retreated, producing a rise in sea levels. The tundra that had covered much of Europe and Asia was gradually replaced by forests and grasslands, bringing new plants and animals. Between 17,000 and 14,000 years ago, there were episodic warm periods followed by a final prolonged cold interval between 13,000 and 11,500 years ago. Then the climate stabilized, with warmer, wetter weather that was more suitable for agriculture.[2]

As noted in the last chapter, as the earth warmed, people began coming together in larger groups than had previously been the case. For hundreds of thousands of years, hominins had mostly lived in small bands of hunter-gatherers. Beginning about 18,000 years ago, however, there are suggestions that bands of hunter-gatherers joined together at certain times of the year to hunt cooperatively. In France and Spain, such aggregation sites have been identified with what appear to be permanent shelters. The bands presumably hunted separately during some seasons but came together and hunted cooperatively during other seasons. This was probably the first time in history in which large numbers of hominins gathered together on a regular basis, and it would have remarkable consequences. One of these was Göbekli Tepe.

"THE FIRST HUMAN-BUILT HOLY PLACE"

Discovered in 1995, Göbekli Tepe sits on a hilltop near Urfa, in southeastern Turkey. Its construction began 11,500 years ago, at about the same time the last caves were being painted in southern Europe. It is therefore 7,000 years older than Stonehenge.

Spread over 22 acres, Göbekli Tepe consists of 20 enclosures, some with terrazzo or stone floors, limestone pillars, and stone benches. The enclosures may originally have been covered with roofing. The pillars, of which there are approximately 200, are up to 18 feet in height and weigh up to 15 tons each. They are of great interest to archeologists because they are T-shaped; some have carved arms and hands on their sides, a belt with a buckle at the midsection, a loincloth below the belt, and even a necklace near the top of one pillar. Thus, it seems clear that the pillars represented anthropomorphic beings of some kind, with the top of the T representing the head. Many of the pillars are decorated with carved animals, especially dangerous animals such as snakes, foxes, vultures, scorpions, spiders, lions, and wild boars. Carved human figures are relatively rare, similar to the situation in the European cave paintings.[3]

Although only a fraction of Göbekli Tepe has yet been excavated, there are already many findings of note in addition to the stone pillars. Among them are several life-size, crudely carved stone human heads. There are also what appear to be carved stone totem poles; one such pole, discovered in 2009 and now on display at the museum in Urfa, is about six feet tall and shows what appears to be a bear holding a human, who is holding another human, who is holding an unidentified animal or possibly giving birth. Large snakes decorate both sides of the totem. The pole is strongly reminiscent of the wooden totem poles carved by the Northwest Coast Indians more than 11,000 years later.

What was the purpose of Göbekli Tepe? So far, no houses, cooking hearths, trash pits, or other evidence of permanent habitation has been found in association with it. On the other hand, thousands of deer, gazelle, and pig bones, as well as stone bowls and goblets, have been found, suggesting that feasting took place there. This idea received support from an analysis of residue contained in drinking vessels excavated from Körtik, a nearby archeological site thought to be approximately contemporaneous with Göbekli Tepe. The preliminary analysis of the residue suggested that the vessels had contained wine.[4]

Although only a few human bones have yet been found at Göbekli Tepe, Klaus Schmidt, the German archeologist who excavated there for almost two decades before his death in 2014, suggested that Göbekli Tepe was "a burial ground or the center of a death cult . . . the first human-built holy place . . . the world's oldest temple." Was the function of the carved dangerous animals to protect the dead? Schmidt also believed that Göbekli Tepe functioned as a pilgrimage site and ceremonial center "for settlements at least 50 kilometers away. . . . This is the cathedral on the hill."[5]

Although it is the largest such center discovered to date, Göbekli Tepe was not the only ceremonial center in southeastern Turkey dating to this period. Hallan Çemi, dated to approximately 12,000 years ago, is another. According to Michael Rosenberg, the archeologist who carried out the excavations, "There is strong evidence for the existence of structures that are not strictly domestic in nature and instead served some kind of public function." Numerous stone bowls, many of them decorated, and sculpted stone pestles are thought to have been associated with public feasting. In addition, one of the public structures "contained a complete aurochs skull that appears to have once hung on its north wall."[6]

Nevali Çori, settled about 10,500 years ago, is only 20 miles from Göbekli Tepe and also had a building with a terrazzo floor, stone benches, and T-shaped pillars. On one of the central pillars, "two bent arms with clasped hands had been carved." Found in the building were "several fragments of huge limestone sculptures," "fragments of the head and body of a strange being with a human head and a bird-like body," several miniature limestone masks, and a totem pole with a bird seeming to grasp two human heads gazing in opposite directions. The best-known finding from Nevali Çori is fragments of a limestone bowl with a carved relief on its outer surface depicting two humans dancing with a large tortoise.[7]

Çayönü, close to Göbekli Tepe and Nevali Çori, was also originally settled about 10,500 years ago, but its most interesting finding is a building constructed later. The building has a terrazzo floor with "parallel, colored stripes of white lime" embedded in the mortar. In this building were found the remains of approximately 450 people. Most of the remains were disarticulated, with piles of long bones and skulls "arranged in north-south lines, facing either east or west." The skulls of aurochs were also included with the human skulls. The building has been called by contemporary writers a "skull house" or "house of the dead."[8]

The "skull house" at Çayönü was in active use for approximately 1,000 years and is thought to have had some ceremonial function. One end of the building contained a "highly polished stone slab" on which was found a four-inch black flint blade. Hemoglobin crystals on the blade were ascertained to have come from aurochs, sheep, and humans, thus making it probable that animals and humans were at least cut up, and possibly sacrificed, in the skull house. Manchester University archeologist Karina Croucher thus concluded that "at Çayönü, the Skull Building was used in a performative way, with events apparently focused on the stone slab."[9]

What are we to make of such findings? At a minimum, we can say that large numbers of people joined together to construct what appear to be ceremonial buildings. It has been estimated, for example, that approximately 500 people would have been needed to transport the largest stone pillars from the stone quarry to Göbekli Tepe. The most remarkable aspect of this, however, is that all this was done before the domestication of plants and animals and the establishment of large settlements were well underway. There is no evidence of the cultivation of grains at Hallan Çemi or Göbekli Tepe, although there are suggestions that pigs were being domesticated at the former.

This raises a question, as Klaus Schmidt noted, "if maybe the 'invention' of agriculture was an epiphenomena of these huge gatherings of hunters and the accompanying work." Thus, the "mass gathering of people over a considerable period of time . . . may have been a catalyst for the domestication of plants." As one journalist summarized Schmidt's thesis: "The construction of a massive temple by a group of foragers is evidence that organized religion could have come before the rise of agriculture and other aspects of civilization." Some support for this thesis has come from DNA studies showing that "the closest known wild ancestors of modern einkorn wheat are found . . . just 60 miles northeast of Göbekli Tepe."[10]

ANCESTOR WORSHIP

What are we to make of Göbekli Tepe, Hallan Çemi, Nevali Çori, and Çayönü? It seems likely that some kind of spiritual activity or ceremony was taking place in those buildings. Are there any clues regarding what spirits might have been honored? The abundant carvings of animals, what appear to be totem poles, and sculpture of a human head with a bird-like body found at Nevali Çori suggest that, like the painted caves, animal spirits were a central focus. But what are we to make of the enormous pillars, some of which have arms and hands, belts, loincloths, and even a necklace? These appear to be some kind of anthropomorphic being. The finding of life-size stone human heads at Göbekli Tepe suggests that human spirits may also have been present. British archeologist Karina Croucher pointed out that the creators of the pillars clearly had an ability to sculpt realistic humans if they had wished to do so, "yet a choice was made to keep these 'beings' ambiguous, with the merging of human form and stone." She concluded that the pillars may "represent a more amorphous category of 'ancestors.'"[11]

If the activities at these ceremonial centers did include the honoring of ancestors, it would not be surprising. Edward Tylor, in *Primitive Culture*, his book about the evolutionary origin of religious thought, claimed that a belief in the existence of spirits of the dead "leads naturally, and it might almost be said inevitably, sooner or later to active reverence and propitiation." James Cox, a professor of religious studies at the University of Edinburgh, said that "indigenous religious beliefs, rituals and social practices focus on ancestors, and hence have an overwhelming emphasis on kinship relations." The emergence of ancestor worship at this time would also be consistent with a recent study of 28 hunter-gatherer societies. It reported that in such societies "belief in an afterlife evolves prior to ancestor worship, and its presence stimulates the subsequent evolution of ancestor worship." By contrast, in such societies, the concept of high gods appears much later.[12]

What are the reasons for honoring one's ancestors? A review of hunter-gatherer cultures suggests that one reason is a belief that deceased ancestors may be of assistance to the living. For example, according to early accounts of the Veddahs of Ceylon, "every near relative becomes a spirit after death, who watches over the welfare of those who are left behind." Such spirits "will appear to them in dreams and tell them where to hunt." The Veddah religion was thus characterized as "essentially a cult of the dead" in which "the propitiation of the spirits of dead relatives . . . is at once its most obvious and important feature." Similarly, among the Siriono nomads of Bolivia, "when a man has had a long streak of ill luck in hunting he may repair to the spot where the bones of an ancestor—one who has been a great hunter—are buried and ask him to change his luck and to tell him where to go in quest of game." A belief that the dead can communicate with and help the living is not unusual even today. A survey conducted in 2009 in the United States, for example, reported that 30 percent of Americans said that they had "been in touch with someone who has already died." It is not uncommon to hear stories such as that of the survivor of an automobile accident, who had put on his seat belt just minutes before a crash, being convinced that his recently deceased grandmother, with whom he had had a close relationship, had warned him to do so.[13]

Hunter-gatherer societies often use formal ceremonies to honor their ancestors, and it may be that similar ceremonies were taking place 11,000 years ago at Göbekli Tepe and surrounding sites. For example, the Ojibwa Indians in Canada held a "Feast of the Dead" every two or three years, at which time "the bones of those who had died during the interim were interred and accompanied by a lavish distribution of foods and other commodities." In some Ojibwa communities, "the dead were represented by wooden figures attached to a string . . . [and] when the drumming began these figures would dance." The Blackfoot Indians also held a "dance for the dead to which their spirits were invited." The Chugach Eskimos held "Feasts of the Dead" each August; relatives "would put gifts into the fire . . . [and] the burnt objects were supposed to go to the deceased." Similarly, the well-known potlatch ceremonies, previously held by the Northwest Coast Indians, were "not an isolated ceremony" but "rather a single episode in a series devoted to the memory of the dead." At these events, "the food served to the guests always included the items that had been most enjoyed by the dead person for whom the potlatch was given."[14]

How common is the honoring of ancestors among hunter-gatherers? It has been described in the great majority of such societies but not in all. According to Geoffrey Parrinder, a professor of comparative religion at Kings College, London, "there is no doubt that ancestral spirits play a very large part in African thought. . . . Many African tribes have no true worship of gods; their place is taken by the ancestors." For example, in Sierra Leone "two distinct groups of ancestors are worshipped . . . those ancestors whose names and feats are known . . . and those who died in the far distant past." On the other hand, the !Kung Bushman of southern Africa have been said to "believe strongly and vividly in the existence of spirits of the dead," but there is no evidence that the !Kung "worship their ancestors or perform any rites of reverence for them." The absence of such evidence, however, should always be treated with caution; as Edward Tylor noted, "it is not always easy to elicit from savages the details of their theology." Thus, given how common it is, ancestor worship would seem to be a likely explanation for the apparent ceremonies that took place at Göbleki Tepe 11,000 years ago.[15]

THE DOMESTICATION OF PLANTS AND ANIMALS

Göbekli Tepe is located at the most northern point of an arc widely referred to as the Fertile Crescent. It stretches for almost 1,000 miles from what is now Israel and Palestine through Lebanon, Jordan, Syria, and southeastern Turkey into Iraq and Iran. Because of the climate and favorable growing conditions, this area had an unusual concentration of wild wheat (both emmer and einkorn varieties), rye, barley, peas, lentils, beans, and chickpeas, as well as wild sheep, goats, cattle (aurochs), and pigs (boars); it was thus ideally suited for the advent of agriculture. The Fertile Crescent is believed to be where the agricultural revolution began.[16]

The first gleanings of what would develop into an agricultural revolution have been found in the Fertile Crescent starting around 20,000 years ago. Although most people continued to live seminomadically in small groups, moving with the seasons and herds of game animals, as noted earlier a few people began spending longer periods in specific settlements. At an unusually well-preserved settlement site in Israel, there is evidence dating to 19,000 years ago that people "were exploiting cereals such as wild barley and wild emmer wheat," as well as collecting wild olives, almonds, pistachios, and grapes. This does not mean they were cultivating the cereals, but merely cutting and using them where they were already growing wild. At a settlement in Jordan "occupied by hunter-gatherers between 20,000 and 10,000 years ago, no fewer than 150 species of edible plants have been identified." Such archeological finds, although unusual, suggest that "Levantine hunter-gatherer groups used wild cereals as food thousands of years before the advent of agriculture." As anthropologist Douglas Kennett succinctly noted: "Agriculture was not a revolution. People were messing about with plants for a very long time."[17]

Given the intelligence of modern *Homo sapiens*, it was inevitable that some people would have noticed that new plants appeared where last year's seeds had been cast aside. This would logically have led to an intentional planting of seeds and then to a selection of seeds from the best plants for future planting. In this way, the intentional cultivation of plants took hold, and agriculture was born. There is evidence that the intentional cultivation of plants began at multiple sites in the Fertile Crescent between 11,500 and 11,000 years ago, at approximately the same time that Göbekli Tepe was being built. These include sites in Israel, northern Syria, southeastern Turkey, and northern Iraq and in the Zagros Mountains in Iran. Since there is abundant evidence of the trade of obsidian, marine shells, bitumen, ochre, and other items among these centers, it is likely that information regarding the cultivation of plants was also being exchanged. At these sites, sickle blades have been found "that show a pattern of wear characteristic of harvesting cereals, rushes, and reeds." Other tools associated with food preparation have also been uncovered, such as milling stones, mortars, and pestles.[18]

Over time people discovered that grasses such as wheat and barley had other uses. If ground to flour, mixed with water, and baked, the grasses could be made into bread. Adding yeast, a naturally occurring fungus, leavened the bread. At some inevitable point in time, barley gruel and yeast were probably accidentally allowed to stand, and the result was a fermented beverage we call beer. Archeologist Patrick McGovern at the University of Pennsylvania has suggested that not only beer but also wine was discovered early in the Fertile Crescent. The Taurus, Zagros, and Caucasus mountains there are regarded as the origin of Eurasian grapes, since this is "where the species shows its greatest genetic variation and consequently where it might have been first domesticated." According to this theory, people would have been collecting wild grapes and storing them in containers. The grape skins would have contained natural yeast, and if left to stand, the grapes would have slowly fermented "into a low-alcohol wine—a kind of Stone Age Beaujolais Nouveau." Then, according to McGovern, "one of the more daring members of the human clan takes a tentative taste of the concoction," reports its pleasing effects to his companions, and invites them to partake. This would have been the world's first wine tasting; there would be no turning back.[19]

Although the initial discovery of beer is often treated humorously in historical accounts, it may, in fact, have played an important role in the domestication of plants. For more than half a century, occasional scholars have suggested that "the domestication of cereals was for the purposes of brewing beer rather than for basic subsistence purposes," sometimes referred to as the "beer before bread" hypothesis. Archeologist Brian Hayden and his colleagues at Simon Fraser University undertook a detailed examination of this thesis, including determining the technology that would have been involved at that time in beer making. They noted that the making of beer was extremely unlikely to have been done in traditional hunter-gatherer societies and most likely began when *Homo sapiens* became semisedentary. They also noted that "the cereals that were first domesticated (rye, einkorn, emmer, and barley) have been shown to be suitable for brewing" and that "there do not seem to have been any significant technological impediments or constraints to the development of brewing in the Late Epipaleolithic [about 12,000 years ago]." Hayden et al. also suggested that the initial brewing of beer was primarily associated with feasting:

> Brewing beer is a laborious and time-consuming process that requires surplus amounts of cereals and control over significant labor. . . . It is not something which is undertaken by families of meager means nor by individuals for frivolous purposes such as ephemeral personal whims or pleasures. The ethnographic literature makes it very clear that brewing beer is done by those with surpluses almost exclusively for special occasions that are socially significant. It is for this reason

that brewing is an essential constituent of feasts in most areas of the traditional world.[20]

As previously noted, there is evidence of feasting between 12,000 and 11,000 years ago at sites such as Göbekli Tepe and Hallan Çemi and there are suggestions of residual wine in the drinking vessels at nearby Körtik. Insofar as such feasting was being done to honor dead ancestors or other spirits, beer and wine, which are well known to stimulate "the mystical faculties of human nature," may have been used to assist in communicating with such spirits. If this is true, then it is possible that beer and wine played a significant role in the early development of religious ideas.[21]

At about the same time that modern *Homo sapiens* began cultivating plants, they also began domesticating animals. The sequence of these two events has been debated, but they probably influenced each other. For example, the parts of domesticated plants that are not used by humans could have been fed to goats and pigs. Similarly, domesticated cattle, oxen, and horses could have increased the areas under cultivation by pulling plows.

Dogs almost certainly were the first animals domesticated. There are claims that this occurred as early as 32,000 years ago and that domestication occurred more than once. As Dale Guthrie of the University of Alaska noted, "Dogs were likely a revolutionary aid in finding game, holding it at bay, and tracking down wounded game—magnifying hunting success perhaps several times over the Pleistocene norm." By 11,000 years ago, when other animals began to be domesticated, domesticated dogs had become widespread.[22]

Sheep and goats were probably the next animals to be domesticated, and there is evidence that this happened by at least 10,000 years ago. Modern *Homo sapiens* was an astute observer of animal behavior, as their cave paintings attest, and would have noted how wild sheep and goats followed a leader and how newborn animals, if removed from the flock, could be tamed. Both sheep and goats would have made important contributions to the lives of early farmers. As noted by Juliet Clutton-Brock in *Domesticated Animals from Early Times*, "The goat can provide both the primitive peasant farmer and the nomadic pastoralist with all his physical needs, clothing, meat, and milk, as well as bone and sinew for artifacts, tallow for lighting, and dung for fuel and manure." Goatskins can also be used for clothing and as water containers.

Pigs (boars) and cattle (aurochs) are thought to have been domesticated next, although in some parts of the Fertile Crescent there are suggestions that pigs were domesticated before sheep or goats. The domestication of cattle was especially important. They provided meat, milk, butter, and cheese; their hides could be made into clothing, shoes, and shields; their dung could be used as fuel or fertilizer or mixed with straw for building; their fat could be burned; and their horns could be used as weapons.

Cattle could also be used to pull carts, turn wheels to bring water from wells, and thrash grain by walking on it. It is thus not surprising that cattle have been revered, perhaps even worshiped, in many cultures, including some of the earliest cultures in southwest Asia.[23]

Within the Fertile Crescent, the domestication of plants and animals was not a neat, linear process. The Fertile Crescent is approximately 1,000 miles in width, and the agricultural revolution took place over a time period of 5,000 years, from approximately 12,000 to 7,000 years ago. Hunting and gathering continued to be practiced for centuries as agriculture was slowly introduced. As Karina Croucher noted, "Neolithization took place over a span of several thousand years, with variable emphasis from region to region."[24]

FARMING AND PARALLEL EVOLUTION

It is believed that the domestication of plants and animals spread from the Fertile Crescent to adjacent regions but also developed independently in other places in the world. Diffusing westward, farming reached western Turkey by 9,000 years ago and was introduced in southeastern

Europe, especially in the countries we now call Greece and Bulgaria, by 8,000 years ago. It continued slowly westward, reaching central Europe about 7,500 years ago, and Britain by 6,000 years ago. Recent genetic and linguistic studies have confirmed that agriculture was brought to Europe by people whose roots lay in southeastern Turkey and the Fertile Crescent, and there is no evidence that European agriculture developed independently.[25]

Diffusing eastward from the Fertile Crescent, agriculture spread to what is now called Iran and Turkmenistan, then to Pakistan and India. By 7,000 years ago, it had become established in the Indus Valley. In moving in this direction, agriculture was following ancient trade routes that had apparently existed for thousands of years. For example, it was reported that seashells used as personal ornaments found in a Jordanian village more than 19,000 years old had come from the Indian Ocean, suggesting "far-flung social networks" long before the agricultural revolution began. Agriculture also spread south from the Fertile Crescent into Egypt and was widely established there by 7,500 years ago.[26]

In addition to diffusing to adjacent regions from the Fertile Crescent, the domestication of plants and animals also took place independently at several other geographically disparate areas of the world. As summarized by anthropologists Robert Wenke and Deborah Olszewski in *Patterns in Prehistory*: "One of the particularly striking characteristics of the 'agricultural revolution' is that not only was it rapid and widespread, but it also happened independently in different parts of the world at about the same time." Indeed, this fact is regarded as one of the strongest pieces of evidence in support of parallel evolution.[27]

Agriculture is thought to have developed at two places in China, probably independently. There is evidence of pottery making in China dated to 20,000 years ago, probably for use in cooking. Rice was domesticated in the Yangtze River Valley by about 8,900 years ago, and millet on the Yellow River flood plain by about 8,500 years ago. Chickens, goats, sheep, oxen, and pigs were also domesticated in China at an early date; for pigs, it was one of at least six times they were independently domesticated, according to genetic studies. Just as in the Fertile Crescent, there is evidence from the residue on pottery shards of the use of fermented beverages during China's agricultural revolution. At Jiahu, in the Yellow River Valley, the residue has been dated to 9,000 years ago and identified by Patrick McGovern as "a complex beverage consisting of a grape and hawthorn-fruit wine, honey mead, and rice beer."[28]

Another area where agriculture was independently developed at about this time was in the highlands of Papua New Guinea. Farming began there as early as 10,000 years ago and included taro, pandanus, bananas, yams, and sugar cane. According to archeologist Peter Bellwood of the Australian National University, the early agriculture in Papua New Guinea "qualifies to be considered as true and primary agriculture, albeit not a highly expansive system in the absence of cereals and domesticated animals." The agriculture in Papua New Guinea did not spread to Australia, probably because the Papua New Guinea highlands are remote and physically inaccessible.[29]

Additional centers for the independent development of agriculture include Peru, Mesoamerica, and sub-Saharan Africa. In the highlands of Peru, "the domestication of plants, including potatoes, was well underway" by 7,000 years ago. On the coast, cotton and other plants were being cultivated by 6,000 years ago. Llamas, alpacas, and guinea pigs were domesticated later. In Mesoamerica, stretching from northern Mexico to Guatemala, evidence of the cultivation of squash dates to 10,000 years ago, followed by pumpkins and beans. Maize (corn), which would become the staple food of the Americas, was initially domesticated in central Mexico about 5,500 years ago. Finally, many experts believe that farming began independently in the Sahel region of Africa, just south of the Sahara Desert, with millet, sorghum, rice, and yams being domesticated. There is also reasonably good evidence that cattle were independently domesticated in this region.[30]

THE LIVING AND THE DEAD

The gradual shift from hunting and gathering to farming that took place between 12,000 and 7,000 years ago brought about profound changes in the relationship between the living and the dead. A migratory lifestyle demands that the deceased be buried or otherwise disposed of wherever they happen to die, since carrying dead persons around is obviously impractical. A sedentary lifestyle, by contrast, allows for the burial of the deceased in the vicinity of the living, and thus the gradual accumulation of the bodies of ancestors from preceding generations. It seems likely that at this time deceased ancestors became much more important to the living.

At one level, local burials facilitate the remembrance of one's ancestors; for example, each time one passes the tree beneath which they are buried, they can be thought about. As Karina Croucher noted, "Keeping the dead close to the living may reflect a desire to retain emotional ties with the deceased, as well as aiding the transitional process of mourning." At another level, burying deceased family members in the vicinity of the living has practical implications for land ownership and kin obligations. The land that was farmed by one's ancestors is the same land on which they are buried and the same land that is being farmed by the present generation. According to one summary: "Often the land and the ancestors are intimately connected. Among many African tribes, ancestors are the ultimate owners or proprietors of the land. . . . Among Australian aboriginals, ancestors are thought to be part of the land itself." Such arrangements inevitably led to the idea of land ownership, probably for the first time in history. The idea of land ownership in turn raised questions about who could inherit the land and how it should be divided following the death of the owner. "Under such conditions," it has been noted, "appeal to some authority was required, and lineage ancestors provided a natural source for such authority." In *The Archeology of Death and Burial*, Mike Parker Pearson described this association between land and people during the agricultural revolution:

> Ancestors of specific kin groups were becoming increasingly important for several reasons. Their physical remains served to tie people to the very earth itself, during a time when the seasonal exploitation of that earth, in the planting and harvesting of crops, was becoming a principal feature of people's lives. For such seasonal tasks, mobilization of large enough groups was essential and, in drawing on each other's

labour, people needed to recall and demonstrate the ancestral genealogies which bound the living together.[31]

Human burials during the agricultural revolution were sometimes accompanied by grave goods, and this became a more common practice as the agricultural lifestyle was more firmly established. Most of the grave goods were utilitarian or decorative and were sex-specific. Men, for example, were accompanied by bone tools, sickles, or obsidian blades, useful for harvesting grain in the afterworld. Women were sometimes adorned with "shell and stone beads, bone pendants worn around the neck, waist, and wrist, and necklaces, bracelets, and belts." Karina Croucher noted that children "under the age of about four years [have] the highest number of grave goods," including "small drinking cups" for refreshment in the afterlife.[32]

In addition to utilitarian and decorative grave goods, some people who were buried at this time were also accompanied to the afterworld by animals or parts of animals. Dogs were buried most frequently; this may simply indicate an affectionate relationship between the deceased persons and their dogs, or the dogs may have been buried to help their masters in the afterlife. In addition to dogs, the "remains of deer, gazelle, aurochs, and tortoise [have also been found] within graves." In some regions, fox mandibles accompanied child burials. A concern for the deceased and one's ancestors was thus becoming more prominent in human history at the same time that plants and animals were being domesticated. Ancestors and agriculture were evolving together.[33]

As modern *Homo sapiens* increasingly settled next to their fields, extended families built houses near one another. Such family clusters slowly grew into villages between 11,000 and 10,000 years ago, by which time a village such as Jericho had a population of approximately 2,000 people. Archeological records have confirmed that in these early villages "adjacent houses were related through kinship." Such clustering of people living permanently together was new in human history. It allowed people to collectively exchange ideas about everything, including how to select the best seeds for planting and how to best honor one's ancestors. As Mike Parker Pearson summarized this period: "What we appear to be witnessing in [southwest Asia] between 12,800 and 10,000 BP (before the present), when farming had its origins, is the beginning of a human obsession with the material presence of the dead among the living."[34]

In the early phases of the agricultural revolution, approximately 12,000 to 10,000 years ago, it was a common practice to bury deceased family members directly beneath the floor of the family's house. Recent excavations have also made clear that in some instances the dead person was buried first and then the house was built directly on top of the grave. In all such cases, the dead person remained physically close to the living. Indeed, "in one instance a sub-floor burial had been placed with its head resting on a stone pillow, resulting in the head protruding clearly into the domestic plaster floor above." As Karina Croucher noted, "It appears to have been important to keep the dead physically close to the living; the living lived their lives in the rooms above their buried descendants." It was not until the later phases of the agricultural revolution that it became common to bury the deceased in common areas adjacent to the village in what essentially became the first cemeteries.[35]

SKULL CULTS

During the early and middle phases of the agricultural revolution, it was a common practice to exhume dead bodies weeks or months after death and to remove the skulls. The skulls were then displayed in the family's house or in a common area in the village. According to French archeologist Jacques Cauvin: "Skulls were in effect lined up on the floor of a house along a wall. Lumps of red clay were brought into the house and served as pedestals. They were thus exposed, set out so that they could be seen. . . . This tendency to arrange human skulls like art objects is new."[36]

Some of the skulls were painted. Others were modeled with plaster so as to resemble a human face. When plastered, "a layer of lime, gypsum, or mud plaster would be placed over the face, recreating a 'fleshed' appearance out of the plaster." Eyes were created by inserting seashells, by using "a whiter plaster which makes them stand out," or by outlining them with black eyeliner. Some of the plastered skulls "have marks that indicate tattooing, which suggests that some were made to be distinctive or individualized, perhaps imitating the appearance of individuals during life." Karina Croucher noted that some skulls may also have had "hair, headdresses or wigs," although this organic material has not survived.[37]

At least 90 plastered skulls have been found, widely distributed in southwest Asia and dated to between 10,000 and 8,500 years ago. According to British archeologist Jacquetta Hawkes, "The best skulls are so finely modelled and life-like that they are works of art as well as cult objects." The effect of these skulls on viewers is striking. When the first such plastered skull was uncovered at Jericho, the lead archeologist described his research colleagues as being stunned: "None of us were prepared for the object [plastered skull] . . . produced in the evening." Archeologists at other sites have also described the finding of plastered skulls as "a highly emotive experience": "We are drawn toward faces, and these are, literally, 'faces from the past.' "[38]

The widespread display of human skulls in southwest Asia during the agricultural revolution has been referred to as a "skull cult." Studies of the wear pattern of these skulls suggest that they were not only displayed but also handled by many people. Some archeologists have cited this as definitive evidence of ancestor worship. Mike Parker Pearson suggested that such skulls "are representations of dead people who were formerly alive and they constitute an embodiment of how the living perceived their ancestral dead." Karina Croucher similarly noted that "death did not mark the end of the body's engagement with the living world, but rather marked a new phase of activity and interaction with the living. . . . They may have been considered active members of the household beyond death, perhaps seen to influence decisions, and play an active role in the events and consequences of the living. . . . The skulls signify the continuing role of the dead in the lives of the living." Indeed, a skull so displayed in a home could quite literally be regarded as the head of the household.[39]

Also of interest in southwest Asia are statues and masks, the majority of which have been dated to the latter phases of the agricultural revolution. The statues vary from small human figurines, such as seated women or standing men in a group, to human figures more than three feet tall with painted facial features and seashells for eyes. For example, at Ain Ghazal, a Jordanian village dated to between 9,250 and 8,000 years ago, 13 36-inch full body statues and 12 18-inch busts were found and are generally regarded by archeologists as "ancestral portraits." At another archeological site, 665 figurines were found. Few in number but of great interest are limestone masks, with accentuated holes for eyes, a small nose, and a mouth that on some masks includes teeth. Twelve such masks, each different in appearance, were found in one region of Israel and are dated to about 9,000 years ago. Some masks were also perforated along their edges, suggesting that they may have been tied around a human head or possibly around a plastered skull. One mask "is made of limestone with red limonitic inclusions [giving] the mask the appearance of a human face covered with blood." This mask also includes small lumps of asphalt with pieces of hair attached, suggesting that it had hair attached in its original state.[40]

What is the meaning of these enigmatic statues and masks? Some archeologists have suggested that "plastered skulls, statues, and masks [form] a common theme" and are interconnected. The frequent finding of them together at the same archeological sites supports the possibility that all are associated with ancestor veneration in some form. Given how masks have been used in many other cultures over the centuries, it is possible that the wearer of the mask represented the deceased, perhaps in a public ceremony.[41]

Kfar Hahoresh, a village in Israel dated to approximately the same period as Ain Ghazal, is unusual in having no evidence of people living there. Instead, it is "thought to have been primarily a place of burial and treatment of the dead." A headless human skeleton was found there, overlying a pit containing the bones of eight wild cattle. Nearby "was an oval arrangement of 15 human mandibles and other remains that included a possible depiction of an animal outlined using human and animal bones." One grave yielded "a headless gazelle carcass with a plastered human skull." Researchers have speculated that this village was a "regional funerary and cult center that served the surrounding villages."[42]

Çatalhöyük, a village in central Turkey dated to the period between 9,000 and 8,000 years ago, also illustrates the spiritual concerns of its inhabitants at this time. It had approximately 5,000 people, who grew three kinds of wheat, barley, and a variety of vegetables, collected wild fruits and nuts, and kept sheep and goats. According to Jacquetta Hawkes: "Among the population were highly skilled wood workers, textile and basket weavers, stone polishers and potters." Obsidian, a black volcanic glass used to make razor-sharp blades for tools and weapons, was mined nearby and traded as far as Syria, Lebanon, and Cypress; flint, wood, and other raw materials were received in return. This was part of an extensive trade network and, according to science writer Michael Balter, "trade may have been a key to Çatalhöyük's wealth."[43]

Death was clearly a major preoccupation for people in Çatalhöyük. Over five hundred burials have been unearthed from the small portion of the village that has been excavated, mostly buried in the floors of the homes; each house has averaged eight burials, but the number ranges from none to sixty-four. Most of the skeletons have been found intact, and one plastered skull has been found, its features painted with red ocher. Grave goods unearthed at Çatalhöyük have included stone mace heads and bone-handled daggers in men's graves and beads and shell necklaces, bracelets, pendants, and copper and bone rings in women's graves. The most unusual grave goods have been "circular mirrors of obsidian set in neat plaster backs," the earliest mirrors known. These presumably were put in the grave to allow the dead to gaze upon themselves in the afterworld.[44]

One of the most interesting findings in Çatalhöyük has been approximately 40 structures referred to by archeologists as "shrines," "history houses," or "cult centers." These structures often include "intricate arrangements of cattle skulls and horns" and plastered walls with paintings and engravings on them. A major theme of the paintings is death, such as paintings of "vultures with vast wings, their hooked and feathered beaks pecking at headless human bodies." In the realm of the grotesque are "carefully modelled female breasts within which are hidden the skulls of corpses of scavenging creatures—fox, weasel and vulture." Aurochs (wild cattle) are also prominent, with a mural of a huge bull in one house covering an entire wall. According to Ian Hodder, the archeologist in charge of excavations, "In many dwellings, one seems hardly able to move without facing some bull's head or painting." Jacques Cauvin noted bulls to be "an almost obsessive theme at Çatalhöyük" and described several frescoes in which "the beast is surrounded by men in movement, armed with bows and throwing sticks." In addition to the bulls, female figures and figurines are common. These include a figurine of a seated woman, flanked by two leopards, with a round object between her legs. This has been interpreted by some as a woman giving birth and gave rise among some contemporary archeologists to the idea of a Çatalhöyük mother-goddess. Others have suggested that the round object between the woman's legs is a human skull.[45]

What is the meaning of these "shrines" at Çatalhöyük? Since they were scattered throughout the town, they have been called "kinship-cult centers" for "venerating ancestors, . . . loci for venerating activity over a long period of time." Like the plastered skulls, figurines, and masks, the shrines were probably associated with ancestor worship. What would have been asked of the ancestors? Based on what is known about ancestor worship in contemporary agricultural societies, there would have been requests for good rains, abundant harvests, and an increased fertility of the land. Ancestors were probably also asked to ensure the fertility of domesticated animals and perhaps women as well. On the other side of the ledger, people would also have asked their ancestors for protection against drought, storms, and other natural disasters, diseases, and, above all, death. Thus, the core concerns of people 12,000 to 7,000 years ago for which they sought the assistance of their ancestors would have been the elementary issues of life and death.[46]

It thus seems likely that during the agricultural revolution, the worship of ancestors became increasingly important in the Fertile Crescent and adjacent southwest Asia. Since the domestication of plants and animals took place in several other parts of the world independently, did the worship of ancestors also accompany the development of agriculture in those places as well?

In China, the veneration of ancestors does appear to coincide with the development of agriculture. At Jiahu, on the Yellow River flood plain, graves have been uncovered dating to 9,000 years ago, at which time millet and rice were first being cultivated. In some of the graves, "the head of the deceased was carefully removed . . . and replaced by six or eight pairs of whole tortoise shells." Some of the shells contained up to hundreds of "small, round white and black pebbles." Other grave goods included utilitarian items such as awls and millstones, jewelry made of jade and turquoise, and bone flutes thought to be among the earliest musical instruments in China.[47]

The archeological site at Jiahu, as noted previously, is also known for having the first known evidence of Chinese winemaking. According to Patrick McGovern, the wine was probably used at feasts for the deceased, where a descendant of the deceased was appointed to communicate with the ancestors. After fasting for seven days, the appointed person drank the equivalent of two bottles of modern grape wine and then commenced communicating. As described in a later Chinese ode:

> The rituals are completed;
> The bells and drums have sounded
> The pious descendant goes to his place,
> the officiating invoker makes his announcement:
> "The spirits are all drunk."[48]

Given the paucity of archeological research and absence of written records, little information is available on the history of ancestor worship in other areas where agriculture was first developed. It is clear that burials with grave goods were taking place at the same time that plants and animals were being domesticated, but it is not possible to know how the deceased were regarded by the living. Burials in Peru have been dated to 8,000 years ago; in some areas intact corpses were buried, while in other areas they were interred as "disarticulated bones jumbled together sometime after the flesh had decomposed or been removed," similar to what was found in southwest Asia. In the Sahel region of Africa where farming began, multiple burials have been found, some dated to 9,500 years ago; grave goods included pots and beads made from eggshells. Early farming in Pakistan's Indus Valley was accompanied by human burials containing exotic grave goods, such as turquoise, lapis lazuli, and conch shells, some of which had come from more than 300 miles away. And when farming began in Egypt 6,500 years ago, the "evidence of funerary goods from tombs of the Badarian period onward indicated a very early belief in life after death." Ancestor worship would become an obsession in Egypt, as will be discussed in the next chapter.[49]

THE EARLIEST GODS

The agricultural revolution, which took place in several parts of the world between 11,000 and 7,000 years ago, resulted in the domestication of plants and animals. As we have seen, it was also accompanied by a revolution in the relationship between the living and the dead that resulted in the domestication of ancestor spirits. Although the latter revolution has been less chronicled, both revolutions would profoundly affect the future development of modern *Homo sapiens*. Agriculture and ancestor worship developed together, the former for sustenance, the latter for succor.

It seems likely that one of the consequences of the latter revolution was the emergence of the first gods, using "gods" in the more restrictive sense, as defined previously. This may have happened between 7,000 and 8,000 years ago, perhaps earlier. Before the gods could emerge, however, two things had to happen. First, some of the spirits had to become very powerful. We can imagine how this might have happened. For example, a man who had been an excellent farmer was honored by his ancestors after his death. His spirit was prayed to and gifts left at a tree on his land at the time seeds were being sown. If good harvests followed over several generations, he might have become known as a powerful harvest spirit. Similarly, a man who had been a great warrior was honored after his death and his spirit invoked to lead his people in battle. If military successes followed over several generations, he might have become known as a powerful warrior spirit. Spirits of nature, such as rain, could also have been elevated in status if, over several generations, the sacrifice of a goat or sheep had resulted in abundant rain. This is, of course, not a new idea; 2,300 years ago, the Greek philosopher Euhemeros of Macedonia said that "gods were originally human rulers who were gradually deified by their subjects." In the nineteenth century, English sociologist Herbert Spencer suggested that "all gods were ancestors, founders of tribes, war chiefs famed for strength and bravery, medicine men of great repute . . . ancestor worship is the root of every religion." Similarly, Edward Tylor noted that some ancestral spirits may "ascend to the rank of deities."[50]

Studies of primitive societies have demonstrated that there is often a continuum of spirits and deities. On one end of the continuum are ancestor spirits of parents and grandparents. More powerful spirits may represent ancestors who died many generations ago; still more powerful is the ancestor who is regarded as the first member of the tribe. Deities similarly range from gods with anthropomorphic characteristics who may be personal gods for a particular group or tribe, to higher and even remote gods who created the world but then have little continuing involvement with it. As one moves up the continuum from spirits to deities, one acquires more supernatural powers. The line dividing the most powerful human spirits from the least powerful of the deities is imperceptible, similar to the line between twilight and dusk. Many researchers have struggled with this problem; for example, in his study of Australian Aborigines, anthropologist Herbert Basedow noted that "it is difficult at times to distinguish between an original spirit ancestor and a deity."[51]

An unusual opportunity to witness a society with a variety of spirits, but apparently without any higher gods, was the discovery of indigenous groups inhabiting the highlands of Papua New Guinea. Although these rugged mountain valleys had originally been settled by modern *Homo sapiens* 40,000 years ago, and agriculture had developed there approximately 10,000 years ago, these groups were unknown to the outside world until the 1930s, when gold- seeking Australian adventurers arrived. What the Australians found were multiple tribes of stone-age farmers living in small villages with well-developed beliefs about an afterlife and about ancestor spirits. Following their discovery, anthropologists studied the tribes and recorded what the people had believed at the time of their first contact with the Australians.

Although the Papua New Guineans were divided into multiple tribal and linguistic groups, all the people encountered by the Australians believed that their strange, white visitors were returning ancestral spirits. They thought that "they were like people you see in a dream . . . spirit people coming openly, in plain sight." One man "wondered if they had come from the sky [or] from under the ground." Another group "speculated that the pale creatures were ghosts . . . come back from the land of the dead to find their relatives." Telenge, a young man of about 18 at the time the Australians arrived, was said to have recalled: "Their skins were so pale they seemed to glow. . . . The only creatures Telenge knew of who were said to have pale skins were ghosts or powerful spirits. These creatures then must be dama [spirits], a conclusion also reached by other men who gazed in amazement from other parts of the garden." One group of people decided that the Australians were not spirits, however, after observing them defecating, which seemed inconsistent with having a supernatural status.[52]

At the time they were initially contacted, the tribes in Papua New Guinea had an elaborate cosmology that included benevolent spirits and malevolent spirits. Most of the benevolent spirits were ancestor spirits that were thought to intervene in human affairs. Most malevolent spirits were nonhuman in origin and were responsible for sickness and death.

Particular spirits were associated with specific localities, and it was said that "spirit people often take the form of birds and call with their voices." Spirits could also take the form of other animals, including wild pigs or pythons. Some tribes had shrines where the spirits could be honored and where "periodic sacrifices of pork were made." There were also formal tege ceremonies, "which aimed to appease the ancestors and placate the dama spirits through gifts and sacrifices of pigs." Some tribes also used masks at ceremonies honoring the ancestors, with the masked person representing the deceased person. Perhaps this is how the limestone masks had been used in southwest Asia 9,000 years earlier.[53]

The second thing that had to happen before the higher gods could emerge was the coming together of a critical mass of people. Groups of hunter-gatherers, which usually numbered fewer than 100 people, may have honored ancestor spirits and spirits of nature, but there would have been little reason to elevate these spirits to deities. As groups of hunter-gatherers came together and settled in villages and towns, however, it would have been necessary to establish a hierarchy among the competing spirits, just as it was necessary to do so among the human leaders of the various hunter-gatherer groups. Out of this hierarchy of spirits emerged the first gods, which were simply high-ranking spirits with advanced degrees conferred by a consensus of the people.

The population increase that took place during the agricultural revolution, fueled by a stable food supply made possible by the domestication of plants and animals, resulted in the coming together of the critical mass of people. When Göbekli Tepe was being used as a ceremonial center 11,000 years ago, the world's population was, according to estimates, approximately five million. When the world's first temple in Mesopotamia was being used to honor a god 6,000 years ago, the world's population, according to estimates, increased to approximately 100 million, and by 2,000 years ago, to 300 million.[54]

A relationship between the size of a population and the type of gods that exist in that population has been clearly established. In 1960 Guy Swanson, a psychologist at the University of California at Berkeley, published a study of gods in 50 "primitive" societies, a sample of George Murdoch's ethnographic database of 556 societies. Swanson reported a significant correlation between societies that were more socially and politically complex (had more "sovereign organizations") and the existence of "high gods" ("a god who rules the world and heavens"). A more recent study reported a highly significant correlation between the size of societies (number of levels of political authority beyond the local community) and the existence of "moralizing gods" ("gods who tell people what they should and should not do"). This relationship was summarized by Azim Shariff, a psychologist at the University of Oregon, in a paper aptly titled "Big Gods Were Made for Big Groups." Sheriff noted that "big gods . . . tend to be relatively recent Holocene innovations and ones that developed only in large, complex societies." The association of "big gods" with big populations is also emphasized in recent books on the god-is-watching-you theory of religion, summarized in chapter 8.[55]

Are there any more precise indicators of when and where the first higher gods emerged in the later stages of the agricultural revolution? Much of the discussion has focused on the enigmatic figurines and statues, some up to three feet tall, that became increasingly common beginning about 10,000 years ago. Some of the statues were originally painted in bright colors and, according to archeologist Jacques Cauvin, "would have been striking."[56]

It has been widely debated whether these figurines and statues represent ancestors or deities. Those who argue that they represent ancestors point to the fact that the appearance of each is different and thus they do not appear to be attempts to convey the image of a single deity. In addition, many of the figurines resemble in facial features the contemporary painted and plastered skulls, which are widely assumed to have been ancestors. Also, the figurines and plastered skulls are often found in association with one another; some researchers have therefore concluded that the figurines were probably "recently deceased female household members" or "abstracted representations of ancestors . . . indicators of an ancestor-based social and religious organization."[57]

Arguments on the other side include the fact that some of the statues have six, rather than five, toes; Jacques Cauvin claimed that this "would seem to confirm [their] supernatural status." The female figurines found at Çatalhöyük have been especially suggested as being deities. James Mellaart, the British archeologist who first excavated the site, claimed that the female figurines represented a mother goddess. "To Mellaart the notion that Neolithic farmers would call upon gods and goddesses of agriculture and fertility to give them spiritual guidance and bless their harvests seemed obvious." Jacques Cauvin similarly asserted that the female figurines represented "a supreme being and universal mother, in other words a goddess who crowned a religious system which one could describe as 'female monotheism.'" Such assertions have made Çatalhöyük "the equivalent of Mecca for the Mother Goddess movement" among some women, and each year "goddess worshippers make the pilgrimage to Çatalhöyük."[58]

In recent years, the interpretations of Mellaart and Cauvin have become a minority view. Most contemporary archeologists view the female figurines of 7,000 to 10,000 years ago as indicating an important role for women and possible association with fertility but nothing more. As Ian Hodder pointed out, at Çatalhöyük the female figurines "do not occur in special places":

> They do not occur in burials or in locations which would suggest special importance. Most of the figurines, in fact, have been found in prehistoric refuse dumps. By contrast, the depictions of bulls at Çatalhöyük do appear to be in important places, often at the center of what may have been a shrine. Thus, if the residents of Çatalhöyük had elevated anything to the status of a deity, it seems more likely to have been a bull than a woman.[59]

Until new archeological evidence comes to light, it is probably futile to try to further identify the emergence of the first gods in time or space. The possible time period covers several thousand years and the area under consideration stretches for 2,000 miles, from Iran to Bulgaria. What may have been true in one area would not necessarily have been true in another. In ancient Greece, for example, Asclepius was venerated as the founder of physicians in some parts of the country but worshiped as a god in other parts.[60]

It was only after the invention of writing, and thus the availability of historic records, that we can be absolutely certain that the gods had emerged. This occurred in Mesopotamia about 6,500 years ago, as will be described in the next chapter. Since the gods appear to have been fully developed at that time, it seems likely that the first gods emerged sometime earlier, but it is not yet possible to specify when or where that took place.

THE BRAIN OF THE FIRST FARMERS

Between 40,000 years ago, when modern *Homo sapiens* apparently first developed an autobiographical memory and an ability to project themselves backward and forward in time, and 11,000 years ago, when the first farmers began domesticating plants, is a period of almost 30,000 years. Why didn't modern *Homo sapiens* begin cultivating plants at the same time that they began using memory devices, burying fellow hominins with grave goods, and painting magnificent animals they hoped to successfully hunt in the future? As Dale Guthrie put it, "Why is it that for 30,000 years we see no agriculture, urban life, written language, pottery, refined metals, cloth, or any other of the dynamic panoply of innovation that shaped the lives of most of our Holocene ancestors?"[61]

One explanation, of course, was the climate, which was cold during most of that period and thus unsuited to the development of agriculture. However, this does not explain the warmer intervals that occurred intermittently at about 38,000, 35,000, 29,000, and 15,000 years ago. Why don't we find any evidence of the domestication of plants during these warmer intervals, either in the Fertile Crescent or at the other sites where agriculture developed independently after 11,000 years ago?

One possible explanation is that although the brain of modern *Homo sapiens* had developed autobiographical memory, it had not yet fully developed another critical faculty needed for the cultivation of crops and the domestication of animals. This faculty is planning, which is not the same as the ability to remember the past and project oneself into the future. An autobiographical memory is a necessary prerequisite for the ability to plan but is not planning itself.

The part of the human brain regarded as the most important planning center is the lateral prefrontal cortex, as seen in figure 6.1 Whereas the medial prefrontal cortex develops earlier in hominin evolution and plays a critical role in the development of self-awareness, an awareness of others, and introspection, the lateral prefrontal cortex plays a relatively minor role in the acquisition of those cognitive skills. By contrast, the main tasks of the lateral prefrontal cortex are planning, reasoning, problem solving, and maintaining mental flexibility; these tasks are often referred to as the executive functions of the brain. As summarized by one researcher: "Having an unusually large lateral prefrontal cortex has made humans exceptionally capable of doing things that are 'unconventional' in the sense of representing novel solutions to behavior problems."[62]

It is known that damage to the lateral prefrontal cortex may seriously impair a person's ability to plan and reason. The planning and reasoning functions of the lateral prefrontal cortex can be tested using neuropsychological tests. One such test, the Tower of Hanoi, tests the person's ability to plan for the future. Another, the Wisconsin Card Sort, tests the person's ability to change plans as circumstances change. These are the kinds of cognitive abilities that would have been essential for early farmers as they planned their crops and the management of their animals. It thus seems likely that modern *Homo sapiens* 11,000 years ago would have done better on such tests than their predecessors would have

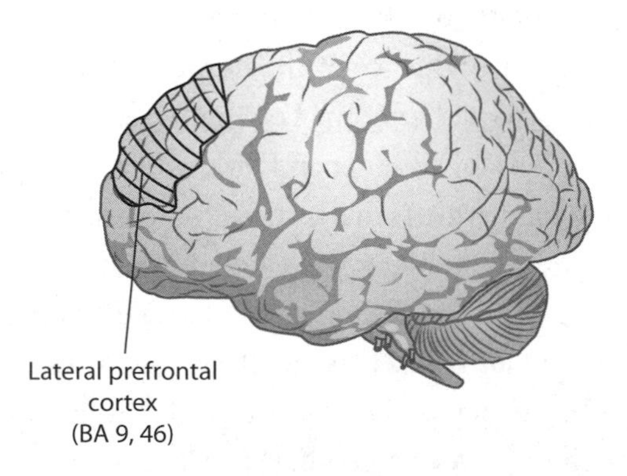

FIGURE 6.1 Modern *Homo Sapiens*: a spiritual self.

40,000 years ago. Those individuals who had the best-developed lateral prefrontal cortex, and thus the best executive brain function, would have been more successful and more likely to pass on their genes.

The fact that the lateral prefrontal cortex is one of the last brain areas to have fully developed in *Homo sapiens* supports this hypothesis. Paul Emil Flechsig, who ranked 45 brain areas by their degree of myelination at birth, ranked the lateral prefrontal cortex among the very last "terminal zones," as he called them. Similarly, neuroimaging studies of gray matter in children's brains have reported that "in the frontal cortex, the dorsolateral prefrontal cortex matures last" and does not reach its full maturation until the person is in their early twenties; this suggests that it developed very recently. When examined under a microscope, the lateral prefrontal cortex also has a different cellular appearance than the remainder of the prefrontal cortex, suggesting that it developed differently. And when the lateral prefrontal cortices from humans and chimpanzees are compared, that of the human is almost twice as large as would be expected. Such observations have led researchers to conclude that this brain area is probably unique in primates and especially well developed in humans. In the opinion of Todd Preuss, one of the leading researchers on brain development: "On present evidence, then, there are good grounds for concluding that dorsolateral prefrontal cortex is in fact one of the distinctive features of the primate brain. In addition, there is evidence that this region underwent extensive modification during primate history."[63]

Accompanying the continuing development of the lateral prefrontal cortex would have been the continuing development of the superior longitudinal fasciculus, the large white matter tract that connects the prefrontal cortex to the parietal and temporal lobes. As previously noted, the superior longitudinal fasciculus is one of the white matter tracts that develop very slowly in humans, indicating that it is a comparatively recent addition in human evolution. A study that compared the gray matter and white matter in the prefrontal cortex in humans and other primates reported that the differences in white matter, the connecting tracts, were much greater than the differences in gray matter, the neurons. Thus, when we ask why *Homo sapiens* did not begin cultivating plants during the climactically warmer intervals 20,000 or 30,000 years ago, the answer may be that there was not yet a sufficient number of connections between the prefrontal cortex and other brain areas. By 11,000 years ago, these connections had developed, allowing for not only the cultivation of plants but the cultivation of the spiritual self as well.[64]

By about 7,000 years ago, it seems likely that the lateral prefrontal cortex and white matter connecting tracts had more fully developed in *Homo sapiens*, enabling the cognitive processes and behavior that we associate with our modern selves. We are able to cultivate both plants and our spiritual selves. The arrival of the gods initiated a period during which formal religions would develop and preoccupy mankind, a period that continues to the present.

7

GOVERNMENTS AND GODS

A Theistic Self

The greatest mystery is not that we have been flung at random among the profusion of the earth and the galaxies, but that in this prison we can fashion images of ourselves sufficiently powerful to deny our nothingness.

—ANDRÉ MALRAUX, *MAN'S FATE*, 1932

The ultimate arrival of the higher gods should not have been a surprise. As spirits, they had been standing in the wings for thousands of years, practicing their lines, waiting for their call to the world's stage. The people who buried their kin with grave goods at Sungir and Dolní Věstonice 27,000 years ago had definite ideas regarding an afterlife, but there is no evidence that the afterlife had divine overseers. The people who painted animals at Lascaux 17,000 years ago exhibited a reverential appreciation of animals' spirits, but there is no evidence that these spirits were transcendent. The people who gathered at Göbekli Tepe 11,000 years ago may have been worshiping the spirits of their ancestors, but the ancestors had apparently not yet become deities. However, sometime during the next 4,000 years, it seems likely that some of the ancestors were gradually elevated to deities. The gods had finally arrived, and when they came, they came to stay.

MESOPOTAMIA: THE FIRST DOCUMENTED GODS

The first god for which there is written, and thus unequivocal, evidence of his existence was Enki, the Mesopotamian god of water. We know this because a temple dedicated to Enki, dated to about 6,500 years ago, was excavated in Eridu, in what was then Mesopotamia and is now southern Iraq. Mesopotamia and surrounding southwest Asia had undergone a rapid increase in population in the preceding years; one study estimated that the increase had been 50-fold, from 100,000 to five million people between 10,000 and 6,000 years ago. By 5,500 years ago, Mesopotamian cities such as Eridu had populations of 35,000 or more; by 5,000 years ago, Uruk is estimated to have had 50,000 to 80,000 inhabitants and to have been the largest city in the world. Thus, from the very beginning, the higher gods were associated with large populations.[1]

Mesopotamia, generally regarded as the world's first civilization, rose to prominence between 6,500 and 4,300 years ago. Socially and economically it was a complex society. There was a high degree of job specialization that included farmers, overseers, laborers, fishermen, brewers, bakers, merchants, soldiers, artists, architects, scribes, and priests. The core of the economy was trade: textiles, wool, leather, sesame oil, and barley were exported in exchange for copper from Oman, lapis lazuli from Afghanistan, carnelian from Pakistan, seashells and pearls from India, wood from Lebanon, obsidian from central Turkey, and tin, silver, ivory, and slaves from a variety of sources. Trade took place by sea and land and was so important that the Mesopotamians maintained permanent trading stations in other countries to develop and protect their interests. The Mesopotamians are credited with the first use of a plow, a potter's wheel, a chariot, a sailboat, a legal code, and standardized weights and measures. Most important, they had a written language that accounts for why we know so much about them. For the first time in history, there was a permanent record of what *Homo sapiens* was doing and thinking.[2]

The original temple at Eridu was a modest, 45-square-foot room with "one entrance, an altar, and an offering table." When the temple was excavated, archeologists found that "hundreds of fish bones, including the complete skeleton of a sea perch, still lay on the offering table." The temples at Eridu and other Mesopotamian cities were rebuilt multiple times over the years, becoming increasingly larger and more elaborate. The temple at Ur, for example, was accessed by three sets of stairs, each consisting of 100 steps, and was covered with "tens of thousands of small clay cones that had been dipped in different colors . . . [and inserted] in such a way that they formed polychrome triangles, lozenges, zigzags, and other geometrical designs." The temple was so impressive that it is thought to have been the origin for the biblical story of the Tower of Babel. The inner walls of some temples were "painted with frescoes of human and animal figures" and decorated with precious metals and stones, including silver, gold, carnelian, and lapis lazuli.[3]

In each Mesopotamian city, "the temple was the largest, tallest, and most important building in accordance with the theory . . . that the entire city belonged to its main god, to whom it had been assigned on the day the world was created." Just as Enki was the god of Eridu, so was An the god of Erech, Utu the god of Larsa, Enlil the god of Nippur, Inanna the goddess of Uruk, Nanna the god of Ur, Shara the god of Umma, and Ningirsu the god of both Lagash and nearby Girsu. When the god was on earth, it was believed that he or she actually lived in the temple.[4]

What do we know about Enki and the other Mesopotamian gods? Thorkild Jacobsen, a Danish archeologist, did an extensive study of these gods and concluded that "the earliest form of Mesopotamian religion was worship of powers of fertility and yield, of the powers in nature ensuring human survival." Thus, the earliest Mesopotamian gods included Utu, the sun god; Nanna, the moon god; Enlil, the god of wind; and Enki, the god of water. Two major themes were evident: the importance of the fertility of the earth to provide food needed for life, and the fate of people after they died. The themes of life and death were thus interrelated in the earliest known religious thought.[5]

The dedication of the first-known Mesopotamian temple to Enki, the god of water, was consistent with the themes of nature and fertility. Enki was called "the fertilizing sweet waters" and "Lord of the soil." According to a Mesopotamian hymn of praise, the duties of Enki included the following:

> To clear the pure mouths of the Tigris and Euphrates,
> to make verdure plentiful,
> make dense the clouds, grant water in abundance
> to all ploughlands,
> to make the grain lift its head in furrows
> and to make pasture abundant in the desert.

In addition to the fertility of the land, Enki was also responsible for the fertility of animals and people. According to Jacobsen, the Mesopotamian language "does not differentiate semen and water: one word stands for both."[6]

Another early Mesopotamian god, Dumuzi, combined the themes of life and death. On the one hand, Dumuzi was the "god of fertility and crops," especially grain. Dumuzi was also married to Inanna, the goddess of the food storehouse. According to Jacobsen: "That these two powers are wed means that the power for fertility and yield has been captured by the numen of the storehouse," thus ensuring an adequate food supply for the community. As such, Dumuzi and Inanna represent life and protection from starvation.[7]

As the god of grain, Dumuzi was also "the power in the barley, particularly in the beer brewed from it." Dumuzi was assisted by Ninkasi, "a special goddess in charge of beer preparation" whose name means "the lady who fills the mouth." Beer was the most popular drink in Mesopotamia, usually drunk with straws communally by several people sitting around a large shared beer jar, as depicted on a clay seal dated to 5,850 years ago. The fact that some of the earliest Mesopotamian gods had responsibility for the brewing of beer is a measure of its importance. The word *alcohol* is, in fact, Mesopotamian in origin.[8]

Unfortunately, just as spring and summer ultimately die, so did Dumuzi. According to one Mesopotamian text, Dumuzi "was set upon by highwaymen," killed, and taken to the afterworld from which nobody, not even a god, was allowed to permanently escape. Inanna searched for and found her husband, guarded by Ereshkigal, goddess of the afterworld. Inanna negotiated an agreement whereby Dumuzi would be allowed to leave the afterworld for six months each year, during which time the grain would grow and be stored, but Dumuzi then had to return to the afterworld for the other six months. The Dumuzi story was thus intended to explain that cycle of the seasons and was the prototype on which similar stories were founded, such as that of Tammuz in Babylonia, Osiris in Egypt, and Persephone in Greece.[9]

The fact that death figured prominently in the story of Dumuzi is consistent with the importance of death in early Mesopotamian religion. The afterworld was conceived of as a "huge cosmic space below the earth" beyond a river crossed by a ferry and guarded by seven gatekeepers. Ereshkigal was thought to live in a temple made of lapis lazuli, and the denizens of the afterworld were all naked. The deceased were believed to undergo a judgment by Utu, the sun god, and Nanna, the moon god, who decreed the fate of the dead, depending on the life they had led. The gods favored those "who were good parents, good sons, good neighbors, good citizens, and who practiced virtues" such as the following: "To the feeble show kindness, do charitable deeds, render service all your days. . . . Do not say evil things, speak well of people." Mesopotamians buried their dead either beneath the floor of the house or in cemeteries. Most burials included grave goods consisting of personal effects such as jewelry and daggers. Many burials also included a cup, bowl, and jar, which held food and beer for the trip to the afterworld. In the city of Lagash, one burial included "7 jars of beer, 420 flat loaves of bread, 2 measures of grain, 1 garment, 1 head support, and 1 bed."[10]

Another indication of the Mesopotamian concern about death and the afterworld is the narrative poem *The Epic of Gilgamesh*. This is the best known of several Mesopotamian poems that have been passed down and is generally regarded as "the earliest classic of world literature." Gilgamesh was the king of the city of Uruk about 4,700 years ago. When Enkidu, his dearest friend and fellow adventurer died, Gilgamesh realized that he too would die and he became terrified. "Enkidu my brother, whom I loved, the end of mortality has overtaken him. I wept for him seven days and nights till the worm fastened on him. Because of my brother I am afraid of death. . . . Do not let me see the face of death which I dread so much. . . . How can I be silent, how can I rest, when Enkidu whom I love is dust, and I too shall die and be laid in the earth."[11] Gilgamesh then embarked on a quest to find the secret of immortality. His journey took him to the ends of the earth, where a woman told him: "You will never find that life for which you are looking. When the gods created man they allotted to him death, but life they retained in their own keeping." Undeterred, Gilgamesh pressed on and went to the afterworld to find Utnapishtim, the only human to whom the gods had granted immortality. They did so because Utnapishtim, thought to have been the model for the later biblical story of Noah, had built a boat and saved mankind at the time of the Great Flood. Utnapishtim told Gilgamesh: "There is no permanence. . . . From the days of old there is no permanence. The sleeping and the dead, how alike they are, they are like painted death." Gilgamesh finally realized that he could not reverse the decree of the gods and, like Enkidu, he too would die. "Already the thief in the night has hold of my limbs, death inhabits my room; wherever my foot rests, there I find death." He returned to Uruk and resumed his duties as king, older but wiser. Eventually Gilgamesh did indeed die and, as described in the poem, "like a hooked fish he lies stretched on the bed, like a gazelle that is caught in a noose."[12]

THE GODS ACQUIRE POLITICAL AND SOCIAL RESPONSIBILITIES

According to Thorkild Jacobsen's study of the Mesopotamian gods, those associated with nature, life, and death were the "oldest and most original" of the gods. This was the first phase of Mesopotamian religion consisting of "the selection and cultivation for worship of those powers which were important for human survival—powers central to the early economies—and their progressive humanization arising out of a human need for a meaningful relationship with them." These gods dominated Mesopotamian religious thinking from 6,500 years ago until about 5,200 years ago, a period called by historians the Uruk period.[13]

In the Dynastic period that followed, from 5,200 years ago until 4,350 years ago, the nature of Mesopotamian society and the nature of the gods changed. The secular rulers or kings of each city-state became more powerful, usurping some of the power of the temple gods. As the kings were assuming some of the authority of the gods, the gods also assumed some secular authority. Thus, Utu, who had previously been exclusively the sun god, also became the god of justice. Nanna, the moon god, acquired responsibility for cattle. And Ningirsu, the god of thunderstorms and floods, acquired responsibility "as protector and military leader."[14]

In this second phase of Mesopotamian religion, it became increasingly common for kings to assume divine prerogatives. Naram-Sin, who came to the throne approximately 4,200 years ago, even declared himself to be a god. Shulgi, who ruled two centuries later, "was worshipped as a god during and after his lifetime." Confusion about the divine status of Mesopotamian kings has led to confusion about the spectacular royal burials at Ur. The 16 tombs contained an extraordinary array of treasures for use in the afterworld. In one tomb, the deceased had a gold helmet, a silver belt, and a gold dagger in a silver sheath, and was holding a gold bowl. Surrounding him were gold and silver lamps, gold and silver axe heads, and "a substantial collection of jewelry," which may have been "a gift to be presented to underworld deities." In other tombs they found gold, silver, and copper vessels, musical instruments, weapons such as spears, daggers, and harpoons, game boards, and "jewelry made of gold, silver, copper, lapis lazuli, carnelian, agate and shell."[15]

What drew international attention to these finds when they were originally uncovered in the 1920s, however, was the fact that some tombs held as many as 73 human sacrifices. In one tomb a queen's "upper body was covered with beads of gold, silver, lapis lazuli, carnelian, and agate—the remains of a beaded cape." She was accompanied in death by 10 women with lyres and harps facing one another in two rows, 11 men, a chariot, two oxen, and "an enormous array of goods." In another tomb a king was accompanied in death by six soldiers, 57 other men and women, two wagons, six oxen, many weapons, and a large number of animal bones that had probably been food offerings. It appeared that the humans being sacrificed had drunk poison, since most had a small cup next to them. Were these the burials of mortals, divine representatives, or gods themselves? As French archeologist George Roux wrote in Ancient Iraq, "the drama of the Cemetery of Ur remains a mystery."[16]

The Mesopotamian gods and their temples represented "the communal identity of each city." One of the gods' most striking characteristics is that, despite having supernatural powers and immortality, they were conceived of as being "entirely anthropomorphic." Like humans, "they plan and act, eat and drink, marry and raise families, support large households, and are addicted to human passions and weaknesses." Because each god was anthropomorphic, the statue of the god in the temple required food twice daily, as well as clothing and entertainment. The food included bread, fish cakes, and fresh fruit. And drinks included beer and wine, left on the offering table. The gods were dressed "in the best finery the community could afford," and over the years "they accumulated more robes, jewelry and other paraphernalia than they could wear at one time." On religious holidays, of which there were many, the statue was taken out and paraded through the streets and, on special festivals, even taken to other cities to visit other gods. Like human families, many of the gods were thought to be related; thus, the statue of Enlil in Nippur was taken to visit the statue of Enki, who was thought to be his brother, in Eridu.

According to George Roux: "It was the duty of every citizen to send offerings to the temple, to attend the main religious ceremonies, to care for the dead, to pray and make penance, and to observe the innumerable rules and taboos that marked nearly every moment of his life." Samuel Kramer, a University of Pennsylvania linguist and expert on Mesopotamia, similarly wrote that the people "were firmly convinced that man was fashioned of clay and created for one purpose only: to serve the gods by supplying them with food, drink, and shelter so that they might have full leisure for their divine activities." The gods thus dominated life in Mesopotamia.[17]

In addition to dominating the social life of Mesopotamia, the gods and temples also dominated the economic life of the city. The temple owned approximately one-third of the land surrounding the city, where temple personnel cultivated cereals, vegetables, and fruit trees, controlled irrigation, and maintained flocks of sheep and goats and herds of cows. Some temple compounds grew to be enormous in size and included workshops that manufactured textiles, metalwork, leather, and wooden items; one temple in Guabba employed 6,000 workers, mostly women and children. The temple staff coordinated trade with other Mesopotamian cities as well as with other countries. The temple also functioned as a community bank, offering loans to merchants at 33 percent interest rates. According to one text, "It seems as though the merchants formally dedicated some of their profits to the temple in order to re-use them as a kind of inviolable capital." Some temples also assumed responsibility for children "when their families were unable to support them," reflecting "a long-standing tradition whereby the temples gathered under their wing the rejects and misfits of society—orphans, illegitimate children, and perhaps the freaks."[18]

This wide array of social and economic activities necessitated a large temple staff, and Mesopotamian records reflect this. A listing of staff for the temple at Nippur includes a high- priest, lamentation-priest, purification-priest, high priestess, treasurer, accountant, scribe, weaver, stone carver, mat-maker, steward, barber, butler, cowherd, boatman, oil-presser, miller, diviner, and snake-charmer. The last was part of the entertainment function of the temple, some of which had "a whole corps of singers and musicians." The Mesopotamian society was extremely well organized, so that "members of the same profession were divided into highly specialized groups"; for example, fishermen were divided according to whether they fished in fresh water or seawater, and "even the snake-charmers formed a 'corporation' which had its own chief."[19]

THE GODS GO TO WAR

During the second phase of the Mesopotamian state, between 5,200 and 4,350 years ago, fighting among the city-states became increasingly common. During the first phase occasional wars had taken place, but the cities had not been fortified and disputes were usually settled peacefully. In the second phase, however, "enormous city walls . . . ringed every city. . . . The larger cities of the region grew as the village populations sought protection behind their walls."

City-states fielded armies of 1,000 to 10,000 men, and wars were waged with spears, shields, battering rams, and siege towers, "some of which would be pre-assembled and floated downstream." The victorious army would often plunder and destroy the defeated city-state, killing the inhabitants or taking them as slaves, and sometimes even destroying the temple of its major god.[20]

The apparent causes of these wars included attempts by city-states to expand their hegemony, land disputes, and the control of irrigations canals or trade routes. Mesopotamian records, however, rarely mention such causes but rather present the wars as conflicts among the gods. For example, a war between Lagash and Umma, for which good records exist, was apparently caused by a dispute over land that lay between them. Umma invaded the disputed land, thereby inciting Lagash to do battle. Lagash prevailed, resulting in "heaped up piles of . . . bodies in the plain"; the victory was commemorated by a carved stele showing vultures devouring the bodies. The Mesopotamian records described it as a "victory of Ningirsu, the god of Lagash, over Shara, the god of Umma."[21]

Thus, the Mesopotamian gods appear to have been intimately involved in the first wars for which we have written records. According to one account, "armies were accompanied, sometimes even led, by a diviner who submitted the plan of campaign to the scrutiny of the gods."

The victorious cities donated some of the spoils of war to their temples; "not to make a dedication to the appropriate temple would doubtless have constituted hubris." The gods were described as inciting wars and "not infrequently displayed hatred and wrath." For example, the god Enlil "'with frowning forehead' puts 'the people of Kish to death' and crushes 'the houses of Erech into dust.'" The consequences of such wars were well described, as when the city of Ur was sacked:

> In all the streets and roadways bodies lay,
> In open fields that used to fill with dancers,
> the people lay in heaps.
>
> The country's blood now filled its holes,
> like metal in a mold:
> bodies dissolved—like butter left in the sun.[22]

In summary, what can be concluded concerning the gods in Mesopotamia, the world's first civilization, between 6,500 and 4,000 years ago? First, it is evident that the earliest gods had responsibility for the fundamental issues of life and death—ensuring an adequate food supply and the fate of people after death. As the civilization became more complex, the gods acquired political, judicial, and social responsibilities such as enforcing laws and providing shelter for orphan children. The temple of the gods became a center for social services. In addition, the gods were used to justify going to war against other cities with other gods. The Mesopotamian wars between city-states were thus the first-known contests between the gods. At the same time that the gods were becoming partially secularized, the secular authorities—kings in this case—were assuming some divine authority for themselves. Thus did religion and politics, the sacred and the secular, become intertwined from the very beginning.

Finally, it is also of interest to note that the Mesopotamians envisioned the first gods as looking and acting just like themselves, with "the appearances, qualities, defects, and passions of human beings." The ancient Greek Xenophanes also noted the human tendency to anthropomorphize their deities and predicted that if horses and oxen could paint their gods, the "horses would paint the forms of their gods like horses, and oxen like oxen." Baron Montesquieu in eighteenth-century France put it more succinctly: "If triangles had a god, he would have three sides."[23]

It is thus clear that the world's first civilization was firmly built upon a religious foundation. As George Roux noted, the gods and ideas concerning them "played an extraordinary part in the public and private life of the Mesopotamians, modeling their institutions, coloring their works of art and literature, pervading every form of activity from the highest functions of the kings to the day-to-day occupations of their subjects."[24]

GODS IN OTHER EARLY CIVILIZATIONS

The Mesopotamian civilization came to maturity between 6,500 and 4,200 years ago. During those same years, civilizations were developing in at least six other areas of the world. Some of these civilizations were influenced by ideas that came from Mesopotamia, while others developed independently. Unfortunately, written records are available only for the Egyptian civilization and, recorded in later years, a civilization in northern China. Nevertheless, it is useful to briefly examine these civilizations to ascertain whether gods also emerged in them as they did in Mesopotamia. The civilizations are those that developed in Egypt, Pakistan, southeastern Europe, western Europe, China, and Peru.

EGYPT

Because it has extensive written records and imposing monumental architecture, Egypt is regarded as the second most important early civilization after Mesopotamia, from which it acquired writing and many of its ideas. By 7,500 years ago, agriculture was well established in the Nile Valley; the annual flooding of the river resulted in rich farmlands, a surplus of food, and a rising population. By 5,500 years ago, towns such as Naqada and Hierakonpolis in Upper Egypt had populations of 10,000 or more.

Administratively, Egypt was divided into 42 regions, unified under the first pharaoh by 5,100 years ago. Its society was stratified, with slaves, farmers, craftsmen, artists, engineers, administrators, scribes, physicians, priests, and a nobility that included a pharaoh. The economy was centralized, with fixed prices, and the temples were the focus of economic activities. Trade was a major source of its wealth, with the export of grain, linen, papyrus, and finished products and the import of gold from the Sudan, ebony, ivory, and wild animals from Ethiopia, timber from Lebanon, olive oil from Greece, copper and tin from Turkey, and lapis lazuli from Afghanistan. The Egyptians built the first true ships and were proficient in mathematics and medicine.

By the time Egypt was unified 5,100 years ago, temples had been built to honor the gods. Such temples were "not a place of meditation . . . but, instead, a home of the god." As in Mesopotamia, the first gods represented natural forces and were concerned with issues of life and death. These included the sun gods Horus and, later, Ra; a moon god, Thoth; the sky god, Nut; an air god, Shu; and a storm god, Seth. Also like Mesopotamia, many gods acquired secondary secular tasks. For example, the moon god, Thoth, was also responsible for writing, knowledge, calculation, and timekeeping. And, as was the situation in Mesopotamia, the gods in Egypt were regarded anthropomorphically.[25]

Another early god was Amon, who originally was the local fertility god for Upper Egypt. In later years, Amon was regarded as the most important of all the gods and merged with the sun god as Amon-Ra. The most important Egyptian fertility god was Osiris, associated with the flooding of the Nile and good crops. In a replay of the Mesopotamian Dumuzi-Ianna myth, Osiris was married to his sister, Isis, killed by his jealous brother, Seth, and then restored to life by Isis. After that, Osiris became the god of the underworld but also continued to return to earth to bring about good harvests.

The outstanding characteristic of Egyptian religion, however, was its obsession with death. The Greek historian Herodotus called the Egyptians the most "religious" people he had ever encountered and was intrigued by their "incessant and elaborate religious rituals." Classicist Edith Hamilton, who wrote extensively about early civilizations, called Egypt "a splendid empire—and death a foremost preoccupation":

> Countless numbers of human beings for countless numbers of centuries thought of death as that which was nearest and most familiar to them. It is an extraordinary circumstance which could be made credible by nothing less considerable than the immense mass of Egyptian art centered in the dead. To the Egyptian the enduring world of reality was not the one he walked in along the paths of every-day life but the one he should presently go to by the way of death.

In Egyptologist Salma Ikram's view, "Death was part of the journey of life, with death marking a transition or transformation after which life continued in another form, the spiritual rather than the corporeal."[26]

The earliest human burials in Egypt were simple graves in the desert with few grave goods. By 5,500 years ago many of the growing towns had large cemeteries, and burials were becoming more elaborate. Some burial chambers were "lined with brick and, depending on the status and wealth of the deceased, divided into sections for different grave goods." An early tomb at Hierakonpolis was adorned with wall paintings depicting "scenes of fighting, hunting and river travel."[27]

By 5,000 years ago, burials of the nobility and commoners had become separated, with royal cemeteries established at Abydos and Saqqara. One royal burial site was surrounded by walls 404 feet long, 210 feet across, and 36 feet high. At another, "tombs of the king's servants were laid out in orderly rows around the royal burial. . . . It seems that the servants either volunteered to die or were forcibly put to death so that they could accompany the king to the Afterworld."[28]

But the elaboration of burial sites was just beginning. It had become common in Egypt to build a rectangular stone shelter on top of royal graves, presumably as a monument of commemoration. Then, 4,600 years ago, a pharaoh named Zoser elaborated this idea at Saqqara by building a smaller stone shelter on top of the first, then a still smaller stone shelter on top of the second, and so on, until it was five stone shelters, or 200 feet, high, effectively having created the first Egyptian pyramid. Each succeeding pharaoh insisted on having a grander place for burial, thus setting off what became an outburst of pyramid construction. It culminated in the Great Pyramid of Giza, built by pharaoh Khufu 4,500 years ago. This pyramid covers 13 acres, is 481 feet high, and required over two million blocks of limestone, some weighing as much as 15 tons. Justifiably, it was regarded as one of the Seven Wonders of the Ancient World.

What was the thinking behind such extraordinary elaborations of burial places? Fortunately, we have Egyptian written records that allow us to answer this question. Egyptians believed that people continued to live after death but in other forms. One form, the Ka, was identical to the body as it appeared in life. The other form, the Ba, was the person's spirit or soul. After death Osiris, the god of the dead, put the person's heart on a balance and weighed it against the principles of truth, wisdom, righteousness, and cosmic order. If the person had lived a good life, the heart was light and the scales balanced, thus assuring the person everlasting life in the Field of Reeds, the Egyptian term for the afterworld. If the person's heart was heavy with sin, however, the scales would not balance and the person was denied everlasting life.

Because the person's Ka was identical to their earthly form, mummification to preserve the earthly form was important. The science and art of mummification became highly developed in Egypt and is regarded as a hallmark of that civilization. In its most elaborate and expensive form, the mummification process could take three months or longer and was available only to royal and wealthy persons; others had to make do with partial mummification or, if they were poor, none at all.

For complete mummification, the brain was extracted by making a hole at the base of the skull; it was then discarded as unimportant. An incision was next made in the abdomen so that the lungs, liver, stomach, and intestines could be removed. These were thought to be important for life, so they were placed in four Canopic jars, each associated with a specific god who looked after that organ. The Canopic jars were stored with the body. The heart was never removed, because it would be needed by Osiris for weighing on the scales.

The body was then saturated with a dehydrating solution, natron, for 70 days. At the end of that time, the body was desiccated and brittle. It was then carefully bandaged, using prescribed rituals, which took 15 days. If any body part broke off, such as a finger, it had to be replaced by an artificial one made for that occasion. It was very important that the body be whole and look as much as possible like the person when he or she was alive.[29]

In addition to human mummification, animals were also occasionally mummified. In some cases, this was done because the animal was a beloved pet of the deceased. In other cases, it was done to provide the deceased with animals to help them in the afterworld. In still other cases, the animal was mummified because it was thought to be an incarnation of one of the gods. Amun, for example, was thought to sometimes appear as a sheep, Hathor as a cow, and Horus as a falcon.[30]

Since Egyptians believed that life in the afterworld would be similar to life on earth, they made provisions to take their possessions with them. Thus, grave goods in Egypt were abundant and became more so in the later stages of the civilization. According to Carol Andrews's book on the subject, grave goods for a wealthy individual could include "beds, complete with mattress and headrest, chairs and stools with cushions, boxes and chests, kilts, wigs and sandals, walking sticks and staffs of office, wine jars and draw-neck bags, jewellery of every kind, mirrors, stone vessels and fans, gaming boards, tables and stands." In a valiant attempt to prove that you can take it with you, one royal mummy wore 22 bracelets and 27 rings.[31]

The magnitude of Egyptian preparations for the afterlife became widely known in 1922, when the intact tomb of King Tutankhamun was discovered. Paintings on the walls of the tomb included Nut, the sky god, welcoming Tutankhamun to the Field of Reeds and Osiris, the god of death, embracing Tutankhamun. The tomb was overflowing with 40 cans of mummified food, 116 baskets of fruit, 40 jars of wine, inlaid chests filled with clothing, beds, chairs, weapons, and chariots. And in the center, like Russian dolls, was a stunning red quartzite sarcophagus, and within that a gilded coffin, and within that another gilded coffin, and within that a pure gold coffin, and within that Tutankhamun, lying peacefully behind a gold mask.

There were three kinds of grave goods associated with Egyptian burials. The first was food. Egyptians believed that even though the person was dead, their Ka still required sustenance. Thus food was left in the tomb and also placed on an offering table outside the tomb, where it could be accessed by the Ka. Periodically the food outside the tomb was replenished by the deceased person's family. Some food bowls were inscribed with special requests directed to the deceased, asking them for assistance in worldly matters such as sickness or finances. The Ka of royal and wealthy persons ate well, as illustrated by food included in the tomb of a royal princess buried at Saqqara: quail, barley porridge, pigeon stew, grilled fish, a joint of beef and beef ribs, kidneys, bread, wine, fruit, cheese, and cake for dessert. The burial chamber of King Scorpion I, dated to 5,150 years ago, included seven hundred jars of wine, which had been imported from the Jordan Valley.[32]

The second type of grave goods associated with, and perhaps unique to, Egyptian burials was small statues called shabtis. In the earliest days of Egyptian burials, servants of royal and wealthy persons were buried in tombs next to the person they had served, thus allowing them to continue as servants in the afterworld. This practice was later replaced by burying statues of the servants, the shabtis, rather than the servants themselves; it was believed that the shabtis would come to life once they reached the afterworld. Initially only a few shabtis were included, but in later burials the number of shabtis was often 365, a servant for each day of the year. Many tombs included written instructions for the shabtis, reminding them of their duties: "O shabti, if your master is commanded to do any work in the realm of the dead: to prepare the fields, to irrigate the land or to convey sand from east to west; 'Here I am' you shall say."[33] The third type of grave goods associated with Egyptian burials was written instructions for the deceased. These instructions included advice such as how to resume breathing, how to put strength into one's legs, directions for reaching the afterworld, and what to do once you arrive there. One collection of instructions found in many tombs is called *The Book of the Dead*, a sort of Frommer's guidebook to the afterworld. Many of the elaborate painted scenes found on tomb walls are visual instructions for the deceased, even including instructions on how to make beer.

The Egyptian pantheon of gods included many who had responsibilities for the dead. In addition to Osiris, who presided over the afterworld, Hathor, a sky goddess, guided the dead to the afterworld, where they were received by Neith, the mother of all gods. The judgment of the dead, when the heart was weighed on a balance, was done by Osiris and Maat, the god of truth and justice, with Anubis, known as "Lord of the Mummy Wrappings," holding the scale.

In Egypt, even more than in Mesopotamia, the sacred and secular were completely integrated. As anthropologist Bruce Trigger noted, the Egyptians had no word for "religion," since "religion was inseparable from daily life." The gods were regarded as all-powerful, and the pharaoh was the gods' representative on earth. Later pharaohs even claimed to be living gods and were worshiped as such. The priests, who were responsible for the upkeep of the temples, became increasingly important as interpreters of the divine will. Thus, government was merely one aspect of religion, which dominated Egyptian life. To reach the Field of Reeds and share everlasting life with the gods was the goal of every Egyptian. Giant pyramids, such as those at Giza, and massive temples, such as those at Thebes, were visual proof that the present life was but a brief stopover on the road to eternal life.[34]

PAKISTAN

The early civilizations of Egypt and Pakistan had two things in common. Both were influenced in their development by ideas from Mesopotamia. And both had a written language, although the language of the latter has never been deciphered. This civilization flourished between 4,500 and 4,000 years ago and geographically is thought to have been the largest civilization at that time. It is usually referred to as the Indus, or Harappan, civilization, the latter being one of its major cities.

The Harappans are best known for their stunning engineering accomplishments. Mohenjo-daro, the largest city, had about 40,000 people, including potters, weavers, brick masons, goldsmiths, and architects. The streets were laid out in an orderly grid, wells and underground pipes carried water, and some houses "had indoor bathrooms connected by drains to a citywide sewage system. . . . The degree of planning that went into Indus Valley urban sites is unmatched among the earliest civilizations." The Harappans also used standardized weights and measures and traded widely from Afghanistan to Mesopotamia for goods such as gold, copper, lead, lapis lazuli, turquoise, alabaster, and carnelian. The importation of agriculture into the Indus Valley was apparently a direct result of trade with Mesopotamia.[35]

In addition to the Harappan written script, many terracotta seals have been found with pictures of what appear to be deities. Of particular note is a three-faced man with a horned headdress, which has been widely assumed to represent an early version of Shiva, a god who is still worshiped by Hindus. Large numbers of female terracotta figures have also been interpreted as "popular representations of the Great Mother Goddess." Sir Mortimer Wheeler, a British archeologist who carried out some of the definitive studies of the Harappan sites, was impressed with evidence of phallus-worship and speculated that, like Shiva, it had also been passed down to the early Hindus as linga-worship. Other suggestions of religious practice among the Harappans in Mohenjo-daro include small temples and a Great Bath that "was perhaps reserved for ritual washing by groups of priests." At one time it was assumed that most of Harappan culture had been imported from Mesopotamia, a major trading partner, but more recent thinking suggests that many Harappan cultural elements were developed independently.[36]

Harappan burials took place both next to homes and in cemeteries. In some cases the person was cremated and the ashes buried. In graves exhumed to date, grave goods have included pottery, jewelry, axes and other weapons, and headrests. One Harappan denizen was buried with beads made of gold, onyx, jasper, and turquoise; another "went to rest in an elegant coffin made of elm and cedar from the distant Himalayas and rosewood from central India." In two instances horses were buried, apparently next to their owners. It is thus apparent that Harappans had major concerns about the afterlife.[37]

SOUTHEASTERN EUROPE

Between 7,000 and 5,500 years ago, a civilization developed in southeastern Europe, primarily in regions we now call Bulgaria and Romania, that made it "among the most sophisticated and technologically advanced places in the world." It is often referred to as the Old Europe culture. Agricultural villages with solidly built, two-story homes and herds of cattle, sheep, and pigs supported crafts persons, including those "among the most advanced metal artisans in the world."[38]

One of the most ubiquitous findings from this culture is thousands of female figurines made of "clay, marble, bone, copper, and gold." These were extensively studied by Marija Gimbutas, an archeologist at the University of California at Los Angeles who believed the figurines were variations on the "Great Goddess of Life, Death and Regeneration," and she published widely on this theme. She also proposed a pantheon of gods to explain the other figurines. In recent years other archeologist have expressed skepticism of Gimbutas's interpretation, but alternative explanations are still needed. The figurines are indeed enigmatic, many being accompanied by chairs for the figurines to sit upon and some found in clusters stored in pottery jars.[39]

Whatever the proper interpretation of the ubiquitous female figurines is, it is clear that people who lived in this culture had an intense interest in the afterlife. In 1972 in Varna, Bulgaria, an extraordinary Old Europe cemetery was uncovered by workers installing an underground electrical cable. Almost 300 burials dated to 6,500 years ago contained a wide array of grave goods, including the first artifacts made of gold found anywhere in the world. There were gold diadems, scepters, discs, pendants, beads, bracelets, armlets, pectoral plates, weapon handles, and even a gold penis cover. Four of the richest graves contained a total of 2,200 gold objects, together weighing almost 11 pounds. Of special interest were 35 graves that contained no body; some of these, however, included expensive grave goods, and three contained clay masks of human faces, embellished with golden earrings and diadems, carefully placed where the person's head should have been. Could these graves have been for people who died elsewhere and whose body could not be recovered, as in the case of someone who died at sea? Whatever the explanation, the people who buried their dead at Varna clearly had an elaborate social organization and explicit ideas regarding the afterlife.[40]

The gold items found at Varna were the product of sophisticated metalwork. British anthropologist Colin Renfrew, who excavated the Varna cemetery, stated it as follows: "The development of metallurgy is one of the clearest cases in which essentially the same innovations were made repeatedly and independently, in different parts of the world at different times. . . . The smelting of ores to produce copper, or the alloying of copper with tin to produce bronze, are in most cases ultimately the same from the technical point of view wherever they are carried out." Renfrew argued that metallurgy had developed independently at several widely disparate places, including southwest Asia, southeastern Europe (including Varna), southwest Europe in Iberia, China, and the Americas, suggesting parallel evolution.[41]

WESTERN EUROPE

The emergence of gods in western Europe is more ambiguous than in Mesopotamia, Egypt, Pakistan, or southeastern Europe. What is clear is that preoccupation with the afterlife was widespread. In France more than 10,000 years ago, a girl was buried decorated with approximately 1,500 shells and beads made from the teeth of deer and foxes and fish vertebra. Similarly, a young woman, decorated with "70 red deer canines, . . . some of which bore incised geometric designs," was buried "lying under two large limestone slabs supported by five stone pillars." These burials took place about 1,000 years after the last cave paintings were done in this region. In northern Russia a cemetery dated to more than 7,000 years ago had more than 400 graves containing 7,000 grave goods, including jewelry made from perforated animal teeth, figurines of animals and humans, and hunting implements.[42]

Beginning about 8,500 years ago, many early burials in western Europe were done using stones. At sites along the Atlantic coast, from Sweden to Portugal, large stones (megaliths) were fashioned to make human graves. In its simplest form, three or four large stones were covered with a capstone that could weigh up to 90 tons; this table-like grave is called a dolmen. In a more complex form, large stones were used to create a passage leading to a room where the deceased were placed; these are called passage or gallery graves, depending on their shape. The entire structure was then often covered with smaller stones, creating a stone hill called a cairn. France has at least 6,000 such megalithic tombs, Denmark and southern Sweden have over 5,000, and Ireland has over 1,200 of these tombs.[43]

Some of those tombs included a single burial chamber, while others had several. Stones in the passage or burial chamber were sometimes engraved with symbols, the meaning of which is not known. Burial chambers contained anywhere from one to hundreds of bodies. Grave goods were relatively sparse but could include jewelry made of shells, limestone, or a turquoise-like stone; tools and weapons such as flint blades, axes, and arrowheads; and pottery, including vessels originally containing a beverage. In some of the graves, the hands and feet of the deceased were severed and a large stone slab placed on top of the body, perhaps to keep the dead from returning to harass the living. The passage leading to the burial chamber was often closed with large stones, thus creating a sealed tomb.[44]

Some of these megalithic tombs are very impressive. The cairn of Barnenez, overlooking the sea near Morlaix in Brittany, was built 6,500 years ago, at the same time the Mesopotamian temple at Eridu was being built and the cemetery at Varna was in use. It has 11 separate passages and burial chambers, in some of which flint blades, polished stone axes, arrowheads, pottery, and other grave goods were found. The cairn is 230 feet long, 82 feet wide, and 26 feet high, and contains more than

13,000 tons of stones. The exterior stones of the cairn were originally placed to create the appearance of a stepped pyramid, similar to those that would be built in Egypt 2,000 years later. André Malraux, as Minister of Culture for France, called Barnenez a "megalithic Parthenon."[45]

It has been said that the construction of megalithic tombs in western Europe was "closely tied to the settling of a new class of farmers." These farmers, who are thought to have come from southeastern Europe, apparently brought with them their beliefs, and scholars have assumed these beliefs included ancestor worship. Many of the burial chambers were large enough for groups of people to have gathered for communal rites, and the use of torches would have dramatically highlighted the engravings on the stones. The probable spiritual significance of the carved stones has long been recognized; in 1805, observing these megalithic tombs, one writer commented on "the peculiar genius of their religion." Recent excavations have also suggested that "important events or ceremonies took place in front of the entrances to the tombs. . . . Large quantities of shards of pottery have been found there, coming sometimes from ceremonial sorts of pots."[46]

Beginning about 5,000 years ago in western Europe, farming became more widespread. At that time, "larger communities formed, some with apparent fortifications of earthworks and enclosure walls of timber." Megalithic tombs continued to be built, but in addition to the tombs, massive construction projects were undertaken. Among the best examples, Brodgar is the least known and Stonehenge is the best known, but Avebury, 20 miles to the south, is the most complete.[47]

Brodgar is located on the Orkney Islands in the north of Scotland. Although the standing stones, the Ring of Brodgar, have been long known, an elaborate complex of interconnected stone buildings has only recently been discovered. One such building is 80 feet by 60 feet, with a stone roof and butterfly-shaped incisions on the stones. The building is thought to have been a "temple or meeting hall" and is referred to by the lead archeologist as the "cathedral" building.

One nearby burial chamber contains 16,000 human bones mixed with eagle talons. There is also evidence of feasting in the stone buildings, and human figurines have been found. Brodgar is still in the early stages of excavation, so additional clues regarding its religious significance will probably come to light.[48]

Both Stonehenge and Avebury consist of standing stone circles, monumental earthworks, and burial chambers, but the latter's are better preserved. The structures at Avebury were built one mile from an ancient hilltop village. People who lived there cultivated wheat, collected wild fruits and nuts, hunted deer, fox, and rabbits, and kept domesticated sheep, cattle, pigs, and dogs. There is evidence at the village site that butchering of animals took place in the spring and fall, at which time there was almost certainly feasting and celebration. The population of the surrounding region at the time has been estimated to have been about 10,000.[49]

Approximately 4,700 years ago the residents of the Avebury region began building a huge earthen pyramid covering 5.5 acres and standing 130 feet high. Today, it is called Silbury Hill, and seen from the air it is almost perfectly symmetrical. Construction was done using bone and wooden tools, then carrying the dirt in wicker baskets. It is estimated that construction took 18 million man-hours, equivalent to 700 men working for ten years, although the work was carried out over more than 200 years.[50]

While Silbury Hill was still being built, an even more massive construction was undertaken one mile away. An outer earthen bank 18 feet high and inner ditch 30 feet deep were built in a circle almost one mile around. The height of the bank from the bottom of the adjacent ditch was thus 48 feet. Inside the circular earthen bank, 98 large stones, one weighing 65 tons, were erected in a circle. Within this outer stone ring, two smaller stone rings were added, one with 29 and the other with 27 standing stones. Seen from the air, the complex resembles a giant face, with the inner stone circles being the eyes.[51]

The primary purpose of Avebury was clearly religious in nature. This is supported by the burial of an estimated 500 individuals in the ditch. Many of the burials are not entire skeletons but rather disarticulated bones, collections of human skulls, mandibles, and long bones. British archeologists Mark Gillings and Joshua Pollard, who have studied Avebury extensively, speculated that the bones were "selected from mortuary deposits elsewhere and may even have been quite ancient ancestral relics by the time they were deposited" at Avebury. Grave goods are surprisingly rare with these burials; according to Gillings and Pollard, they include "a rather bizarre range of items" such as "a dog mandible, a boar tusk, a piece of burnt bone, an antler fragment." Such items, they speculated, may have been "ritual paraphernalia."[52]

The possibility that Avebury was a massive mortuary complex is supported by a stone- lined road that led from the stone circle past Silbury Hill to a hilltop one and a half miles away. Referred to now as the Sanctuary, it consisted of stone circles surrounding some timbered buildings. Archeologist Aubrey Burl, considered to be the leading British authority on megalithic constructions, viewed the Sanctuary as "a series of mortuary houses for the storage of corpses until desiccation was complete and the bones could be removed to nearby chambered tombs."[53]

The nearby tombs included the West Kennet long barrow, a 330-foot-long earth and stone burial chamber a short walk from the Sanctuary. This is the longest stone burial chamber known anywhere in Europe. It contained the skeletons of 46 individuals, most of which were disarticulated. Crematory remains were found in one chamber, and a row of skulls in another. Such burial chambers are thought to have been used "as temporary housing or storage of the dead, removal [of remains] being as common as deposition." Many of the burials in the long barrow were accompanied by ceramic pots, bowls, and cups, presumably for use in the afterlife.[54]

In addition to having a religious function, some have proposed that Avebury also had an astronomical function. It has been compared to nearby Stonehenge, where the standing stones were aligned to correspond with the rising and setting of the sun at the solstices. An astronomical explanation may also apply to Avebury, although nobody has yet proposed a reasonable one. Astronomical explanations are not exclusive of religious explanations but rather complementary, as has been demonstrated at Stonehenge, which began as a cemetery. Many cultures have combined the worship of sun gods with other gods and ancestors, and this may also have been true at Stonehenge and Avebury.

In the absence of any written record, a definitive understanding of Avebury is likely to permanently elude us. What we can say with certainty, to quote Burl, is that "in the new stone age death and the dead obsessed the living." Sites such as Stonehenge and Avebury suggest "scenes of communities engaging in seasonal ceremonies, making offerings in rites made powerful by the manipulation of the dead." As Mark Gillings and Joshua Pollard summarized it:

> The sheer scale of a monument like Avebury provides the most immediate impression of power. Like the awe-inspiring architecture of a medieval cathedral, the magnitude of both the enclosure and the stone settings evokes a sense of the sublime. Here power over people is perhaps exercised through an exaggerated scale that dominates the human body and generates a perceptual awareness of the colossal labour necessary for its production. Such a scale of work could serve to legitimize the authority, whether worldly or supernatural, that lay behind the creation of Avebury.

It is not possible to know whether or not gods were present at Avebury, but the monumental scale of the construction would be consistent with that possibility. It is perhaps also relevant that the stone circles resemble a giant face when seen from the air, as gods in the heavens would have observed it.[55]

CHINA

Between 5,000 and 4,000 years ago in northern China, the Longshan culture developed. It has been called "China's first civilization" and the start of "one of the most brilliant and complex civilizations of antiquity," one that gave us paper, printing, the magnetic compass, gunpowder, paddle-wheel propulsion, and a writing system different from that of Mesopotamia. Agriculture and trade were well developed, and the population density at this time was probably higher than anywhere in the world. Many villages were surrounded by rammed-earth walls, some as much as 20 feet tall and 30 feet thick, because of ongoing warfare between villages. There is evidence of decapitation and massacre of victims.[56]

One of the hallmarks of Longshan culture was ancestor worship. Communication with ancestors was done by divination using "oracle bones," which were the shoulder blades of oxen, water buffalo, pigs, or sheep. A specific question was posed to a dead ancestor; the bone was then heated until cracks appeared, and the pattern of the cracks was interpreted as the answer given by the ancestor. Bones with evidence of having been used for divination have been found in large numbers throughout northern China. Although writing had not yet begun during the Longshan period, it was introduced during the Shang period, which immediately followed the Longshan period and thus has been used to understand the former. It appears that the dead ancestors were expected to intervene with the gods on behalf of the living. The highest god, Di, could only be influenced by the ancestors of royal persons, but lesser gods could be influenced by the ancestors of lesser persons. Some Chinese scholars believe that Di was originally an ancestor spirit, whereas other scholars think Di was originally a nature deity. In addition to Di, "there were a number of nature deities, river and mountain gods . . . a sun god . . . and various deities as well."[57]

Monumental structures were built in China during this period. At Chengzishan, northeast of Beijing, "a massive temple on a platform" was recently uncovered and dated to approximately 4,300 years ago. The platform is 542 feet wide and 2,955 feet long, said to be "nearly half the size of the National Mall in Washington, D.C." It is still being excavated, and its religious significance is not yet clear. However, in one subterranean room beneath the platform researchers found "a life-size ceramic head of a female with inlaid nephrite-jade eyes," reminiscent of sculptured heads found in southwest Asia 5,000 years earlier. It seems likely that it represented either an ancestor or a goddess.[58]

The Longshan culture was also marked by concerns about death and an afterlife. There were "highly differentiated burials" depending on the person's social status. In graves of the social elite, bodies were interred in wooden coffins and often covered with red cinnabar powder. Some graves were "richly furnished with one hundred to two hundred items, including a red pottery plate printed with dragon design; wooden drum covered with crocodile skin; music stone (ch'ing); drum-like pottery;

wooden table, stand, vessels, and other objects painted in bright colors; jade and stone rings; and whole pig skeletons." The grave of one young man included four pottery vessels, 14 stone and jade implements, 24 jade rings, and 33 jade tubes, called congs. The function of the latter is unknown, but some of them have carved heads of animals or humans. Vessels found in graves from the Shang period have been shown to have been filled with various kinds of beer and wine made at that time. Other elite graves have been found with additional human bones; some have interpreted this as evidence of human sacrifice, perhaps servants sent to the afterworld to continue working for their master.[59]

PERU

Despite the absence of written records, it is clear that coastal Peru had a highly developed civilization between 5,500 and 4,000 years ago. A sophisticated system of irrigation-based agriculture produced beans, squash, guava, pacay, lacuma, and cotton. From the ocean, the inhabitants obtained fish, anchovies, clams, mussels, and even sea lions. The various river valleys lining the northern coast traded with one another, with the Andean highlands, and even with the Amazon basin.[60]

The outstanding feature of coastal Peru during this period was the construction of more than 100 platform mounds, many of which had what appear to have been temples on top. The oldest site identified to date is Sechin Bajo in the Casma River Valley. There, in 2008, archeologists announced the finding of a circular stone plaza 5,500 years old. Archeologists theorized that the plaza "may have been a site for gatherings, perhaps a kind of ceremonial center." Sechin Bajo also has a 53-foot platform pyramid. At nearby Sechin Alto, dated to 3,700 years ago, the platform pyramid was 144 feet high and covered an area the size of 14 football fields; it has been called "probably the largest single construction in the New World during the second millennium BC."[61]

Aspero, at the Pacific shore in the Supe Valley, has been dated to 5,000 years ago. It includes six platform pyramids up to 35 feet in height. In one, called Huaca de los Idolos, "a cache of at least 13 small figurines of unbaked whitish-gray clay was buried between two floors of a small room." In another, called Huaca de los Sacrificios, "an infant less than two months old was buried wearing a shell bead cap, and wrapped in a cotton cloth which was placed inside a blanket." The burial has been interpreted "as a dedicatory offering for the public architecture."[62]

The best-known and most extensively excavated site in early Peru is Caral, a UNESCO World Heritage site in the Supe Valley, approximately 12 miles from Aspero. The total population of Supe Valley at that time has been estimated at approximately 20,000 people. The Caral complex covers 160 acres and includes "six large platform mounds, numerous smaller platform mounds, two sunken circular plazas, an array of residential architecture, and various complexes of platforms and buildings." The largest platform pyramid is 100 feet high with a base the size of four football fields. The oldest finds have been dated to 4,600 years ago.[63]

What appear to be altars have also been identified at Caral. An adult and several child burials, appearing to have been human sacrifices, have been found in the largest pyramid. In another pyramid, whale vertebrae "were found in association with two pacay tree trunks . . . in an evident ceremonial context. The tree trunks were driven into the ground and covered with a woven plant fibre fabric." A variety of musical instruments have been uncovered, including bone whistles, cornets, flutes, panpipes, and rattles, which may have been used in ceremonies. Inhalers have also been found, suggesting possible hallucinogen use.[64]

It is also of interest that on the southern Peruvian coast during this period the dead were being mummified, a practice started in Peru more than 1,000 years before it started in Egypt. Initially, the dead were merely salted and allowed to desiccate in the desert sun. Later, according to University of Florida archeologist Michael Moseley, the Chinchoros people in southern Peru and northern Chile became highly skilled at mummification:

> Chinchoros morticians perfected unusual skills: in disassembling bodies; in removing cerebral and visceral matter; in treating organs, structures, and skin to arrest deterioration; in reassembling the corpse

> components; in implanting cane or wood supports into the vertebral column, arms, and legs; in adding fiber, feather, clay or other fill to body cavities; in applying an exterior coat of clay permitting the sculpting and painting of facial details; and in replacing pelage with wigs and human hair embedded in clay.

In some parts of Peru, mummies were wrapped in brightly colored cotton or wool. Grave goods were not common but could include "tools, food, or even pet monkeys or parrots" for the deceased. Mummification was also practiced in highland Peru, where bodies were placed in stone-lined galleries within burial mounds, sometimes "accompanied by textiles and jewelry" such as "shell discs with engraved birds and a stone disc mosaic with a cat-like face."[65]

Although it seems probable that the Peruvians had deities at this time, nothing is yet known about them. At one site, a carving was found that has been referred to as "the staff god." Dated to 4,600 years ago, it is "a fanged creature with splayed feet, holding a snake and a staff." It appears to be identical to a god worshiped by the Incas more than 3,000 years later.[66]

In summary, although higher gods apparently became manifest to modern *Homo sapiens* sometime before 7,000 years ago, we do not have definitive proof of a belief in their existence until the advent of written records. In Mesopotamia 6,500 years ago, there is such proof in the form of a temple built to honor Enki, the water god. Over the following 2,500 years, gods also emerged definitely in Egypt and China, probably in Pakistan, southeastern Europe, and Peru, and possibly in western Europe. In China and Peru, the emergence of gods was almost certainly independent of the other sites, suggesting parallel evolution, whereas for the others, independent development is less certain.

Thus, 4,500 years ago, Mesopotamians in Uruk, the largest city in the world, worshiped the goddess Inanna in her temple. Egyptians marveled at the Great Pyramid of Giza, built to honor Pharaoh Khufu, the representative of the gods; the pyramid would be the world's tallest manmade structure for 3,800 years. In Pakistan the Harappan civilization was at its peak, and the 40,000 residents of Mohenjo-daro visited what may have been temples for honoring gods. In western Europe, large groups of people gathered at what appear to be ceremonial centers at Brodgar, Stonehenge, and Avebury. At Caral in Peru, large platform mounds, some 100 feet in height, accommodated large crowds for some kind of ceremonial function, probably involving deities. And by 4,300 years ago, a similar platform mound was built in China.

Thus, by about 4,500 years ago, modern *Homo sapiens* was emerging as the theistic hominin, and a belief in gods has continued to be one of our defining characteristics. More effectively than animal spirits or ancestor spirits, the gods provided answers for natural phenomena and philosophical questions for thousands of years. Where does the sun go at night? Why does the moon change shape? Why do the stars move? What causes wind and rain, thunder and lightning, floods and droughts? Where did the world come from? Why am I here? And especially, what will happen to me after I die? The presence of the gods has been enormously comforting as we have continued to dutifully cross the stage of life, going about our daily tasks, yet knowing that Pale Death was waiting in the wings. To have been accompanied on life's journey by the symbolic and monumental props of the gods has been a continuing and reassuring source of solitude. Such props quiet the inner voices that whisper about the inevitable end of life's drama. The Stygian shore beckoned uneasily 4,500 years ago, as it still does today.

THE EMERGENCE OF MAJOR RELIGIONS

The gods themselves, however, are not the end of the story. As we saw in Mesopotamia, once the gods emerged, they were adopted by the government and assumed some judicial, social, economic, and even military responsibilities. The sacred and secular, gods and governments, developed together. French sociologist Émile Durkheim claimed that "nearly all great social institutions were born in religion." British historian Arthur Toynbee similarly asserted that "the great religions are the foundation on which great civilizations rest." The relationship between the gods and governments would thus partly determine the shape of subsequent emerging civilizations.[67]

Between 4,000 and 2,800 years ago, the Mesopotamian city-states fell into disarray and were defeated by the Assyrians. The chief Assyrian god, Ashur, was married to Kishar, and together they gave birth to Anu, the sky god; Ea, the god of water and wisdom; and the gods of the underworld. The Assyrians competed for supremacy in southwest Asia with the Babylonians, whose chief god, Marduk, had originally been a fertility and warrior god. As the chief god, Marduk appointed the sun and moon to their proper places in the sky. The Hittites then became the dominant power in this region 3,431 years ago, following the sack of Babylon. Their chief god was Teshub, the god of storms and battle, who was married to Hepat, the sun goddess. At Yazilikaya, in central Turkey, one can see stone carvings of Teshub and Hepat leading a procession of other Hittite gods and goddesses.

In Egypt, the pharaohs of the New Kingdom extended their hegemony over Nubia in the south, and as far as Syria in the north. This was the peak of the Egyptian empire. The same gods continued to be worshiped except for a 17-year period during the reign of Amenhotep IV. He changed his name to Akhenaten and tried to replace Egypt's traditional polytheism with the monotheistic worship of Ra, the sun god, whom Akhenaten called Aten. This period is often cited as the world's first-known example of monotheistic belief. Following Akhenaten's death, his son, Tutankhamun, and the succeeding pharaohs restored the worship of the traditional Egyptian pantheon.

In Pakistan, the Harappan civilization went into decline, in part because of an incursion by Aryan invaders from Iran and Afghanistan. The Aryans spread into northern India, where between 3,700 and 3,100 years ago they composed the Rig Veda, which later became a cornerstone of both Hinduism and Buddhism. The Rig Veda described many gods, including Indra, a fertility god; Yama, the god of the dead; Agni, the fire god; Varuna, the sky god; and Surya, the sun god, who has a swastika as a symbol.

In southeastern Europe, the Old Europe civilization also went into decline, but other civilizations arose. Chief among these was the Minoans, who established a civilization on Crete that achieved its peak approximately 3,600 years ago. The Minoans had few gods but rather a pantheon of goddesses, including those with responsibility for fertility, the harvest, animals, and the underworld. In Crete, the Minoan civilization was supplanted by the Mycenaeans, who invaded from the Greek mainland. The Mycenaeans had developed their own civilization, which included many gods, including Zeus, Hera, Athena, Poseidon, Hermes, and Dionysus. These gods were adopted by the Greeks several centuries later, when they developed their own religion.

In China, the Shang dynasty united large parts of the Yellow River Valley and north central plains for more than 600 years. During this time, writing was independently invented and the first Chinese cities built. The chief god, Shang Di, was the god of agriculture and controlled the wind, rain, thunder, and lightning.

In Peru, a temple was built 2,940 years ago at Chavin de Huántar, located at 10,400 feet in the Andes. The temple housed the chief deity of the Chavin religion, which dominated central and northern Peru. The deity, referred to as the Lanzón, was a 15-foot-tall white granite figure that stood at the end of a narrow stone corridor. According to Richard Burger, the Yale University archeologist who excavated this temple:

> The deity depicted by the Lanzón is strongly anthropomorphic. Its arms, ears, legs, and the five-digit hands with opposing thumbs are those of a human. . . . The large upper incisors or fangs that emerge from the upturned or snarling mouth of the deity are particularly noteworthy. . . . The eyebrows and hair of the Lanzón are shown as swirling snakes and its headdress consists of a column of fanged feline heads. . . . The restricted access into the Gallery of the Lanzón bespeaks an inaccessible, powerful and dangerous god.[68]

The Chavin temple is of special interest because it has "an elaborate maze of small vents and drains." University of Florida archeologist Michael Moseley has suggested that "by flushing water through the drains, then venting the sound into the chambers and out again the temple could, quite literally be made to roar! If this were the case the ceremonial center would certainly have seemed other-worldly to the devoted multitudes assembled in front of it."[69]

THE AXIAL AGE

Beginning 2,800 years ago, the final phase in the emergence of gods and religions as we know them began. The world had profoundly changed. The five million modern *Homo sapiens* who had existed at the beginning of the agricultural revolution had increased to between 200 and 300 million. By means of economic and military conquests, people were becoming consolidated into increasingly larger political units. For example, in China the Shang and then the Zhou dynasties united large territories and populations. In southwest Asia, the Neo-Assyrian empire ruled over southwest Turkey, Syria, Lebanon, Israel, Palestine, Iraq, Iran, Egypt, and part of Saudi Arabia. That empire would be surpassed by the Persian Empire and then by the empire of Alexander the Great, who ruled over a territory that stretched from Greece to the Himalayas.

Great empires require great gods and great religions. The original gods of natural forces, life, and death that had been adequate for Mesopotamian and Egyptian cities 3,000 years before were no longer adequate for empires spanning millions of people in multiple ethnic groups. Just as governance had to be systematized to cover the new world order, so did the gods and religions, since they are an integral part of such governance. Those doing the governing derived part of their authority from the gods.

Thus was born the "axial age," a 600-year period from 2,800 to 2,200 years ago (800 to 200 BCE). During this period, Confucianism, Hinduism, Buddhism, Zoroastrianism, and Judaism were all born, the latter subsequently giving rise to Christianity and Islam. Together, these religions provide spiritual sustenance to 60 percent of people currently alive. Other religions, such as the religion of ancient Greece, also came into existence during this period but subsequently died out; their gods now reside in museums rather than in temples.[70]

During the axial age, Confucius, Lao Tsu, many authors of the Upanishads, Buddha, Elijah, Second Isaiah, Jeremiah, Ezekiel, Socrates, Plato, and Aristotle all lived. Indeed, the lives of Confucius, Buddha, and Second Isaiah even overlapped in time. The period was designated as the axial age by German philosopher Karl Jaspers, because, he said, it represented an "axis in history." "All the vast developments of which these names are a mere intimation," said Jaspers, "took place in those few centuries, independently and almost simultaneously in China, India and the West." British philosopher John Hick noted that during the axial age, "all the major religious options, constituting the major possible ways of conceiving the ultimate, were identified and established and . . . nothing of comparably novel significance has happened in the religious life of humanity since." French philosopher Eric Weil added that during this period the Judaic and Greek civilizations acquired their distinctive shapes, and "other civilizations, practically without contact with, and certainly not influenced by, our nascent thought systems show astonishingly parallel developments." Karen Armstrong, in *A History of God*, similarly observed that during the axial age "people created new ideologies that have continued to be crucial and formative." "For reasons we do not entirely understand," she added, "all the chief civilizations developed along parallel lines."[71]

In surveying the development of these religions, five aspects of them are noteworthy. First, all of them offered an answer to the problem of death. An inscription on Babylon's Royal Way assured its citizens that "Marduk, my Lord, gives eternal life." This principle was summarized by William James in his classic study of religion a century ago: "The first difference which the existence of a God ought to make would, I imagine, be personal immortality, and nothing else. God is the producer of immortality; and whoever has doubts of immortality is written down as an atheist without further trial." Four hundred years earlier, Martin Luther had stated it similarly: "If you believe in no future life, I would not give a mushroom for your God."[72]

Second, major religions provide other benefits in addition to offering a solution to the dilemma of death. Such benefits include the psychological support that accompanies group membership as well as such benefits as physical protection, social services, and access to jobs or economic advancement. Indeed, the psychological and social benefits of some religions may become so prominent that such benefits may appear to have been the origin of the religions. From a sociological perspective, it may even appear that gods are "far from necessary, where human religion is concerned," as Robert Bellah has argued.[73]

Third, as noted earlier, major religions usually develop in conjunction with the political governance of the people. The sacred and the secular develop hand in hand and are often inseparable. Thus, in Mesopotamia the temples of the gods controlled the workshops and trade on which the economy was built. In addition, the political leaders allied themselves with the gods and, in some cases, claimed to have semidivine or even divine status. Nineteenth-century German leader Otto von Bismarck noted this principle when he observed: "The statesman's task is to hear God's footsteps marching through history, and to try to catch on to His coattails as He marches past."[74]

Fourth, religions are continuously emerging, and the success or failure of each is largely determined by the economic, political, or military success of its adherents. Buddhism and Christianity, for example, became world religions largely because they were initially embraced by, respectively, Ashoka, the emperor of India, and Constantine, the emperor of Rome. Conversely, the Greek religion, although initially a major world religion, did not survive, because following the death of Alexander in 323 BCE, the Greek city-states underwent endless civil wars that weakened them politically and reduced their gods to shadows of their former selves. Then, when the apostle Paul began preaching Christianity to the Greeks, Jesus offered a solution to the problem of death that was significantly more attractive than that offered by Zeus.

Finally, the emergence of new religions occurs primarily by borrowing gods and theology from older religions. For example, among the deities of ancient Greece, Aphrodite, the goddess of love and beauty, is thought to have come from Cyprus and to have been "brought to Greece by sea-traders." The Cypriots, in turn, are thought to have borrowed her from Assyria and Phoenicia, where she was Astarte; Babylonia, where she was Ishtar; and, prior to that, Mesopotamia, where she was Inanna.

Similarly, the Greek figure of Adonis, the handsome man beloved by Aphrodite, had previously been a major deity in Phoenicia, where a large temple was built to him at Byblos. Prior to that he is thought to have been borrowed from Babylonia, where he was Tammuz, and Mesopotamia, where he was Dumuzi. The idea of borrowing gods is not a new one. Herodotus, a Greek traveler and historian, noted 2,400 years ago that "gods in different religious systems and with different names and attributes actually had very similar functions," and he specifically conjectured that "the Persians had borrowed the worship of Aphrodite from the Assyrian cult of Astarte."[75]

Just as gods were often borrowed, so too were religious ideas. For example, the Judeo-Christian religion is thought to have taken its ideas for the creation of man, the Great Flood, and the Tower of Babel from the Mesopotamian religion. Similarly, during their exile in Babylon beginning in 587 BCE, the Israelites were exposed to the Zoroastrian religion, with its all-powerful Ahura Mazda. After the Israelites returned to Judah, the idea of an all-powerful, monotheistic god became prominent in the Old Testament for the first time. Other ideas that may have been borrowed from Zoroastrianism include the concept of a "saoshyant," or savior, who would appear "at intervals in the history of the world when it was in danger of such moral degeneration that it might seem that it had fallen finally to the forces of evil." The final savior would be the one to usher in the Day of Judgment, when "each person's good deeds would be weighed against their evil ones." Zoroastrian followers were also taught that three of the saviors would be born to virgins but fathered by Zoroaster, the founder of this religion.[76]

The axial age was thus the culmination of a remarkable period in the evolution of modern *Homo sapiens*. In a mere 4,000 years, the first gods and civilizations emerged, spread rapidly, and were followed by the formation of all of the world's major religions. Robin Dunbar once noted that "religion is the one phenomenon in which we humans really are different in some qualitative sense from our ape cousins," and asked:

"Why is it that, uniquely in the Animal Kingdom, religion has such a stranglehold over our species?" The answer is that we are not only clever, aware, empathic, and self-reflective; we also have an autobiographical memory that allows us to integrate our past as we contemplate our future. This has made us, in the words of Karen Armstrong, *Homo religiosus*.[77]

The dilemma of death was an inevitable consequence of the evolution of the human brain, but gods and religions have provided us with a solution to this innate and infinite dilemma. In doing so, it has made humans into hybrids—half mortal and half immortal. Ernest Becker captured this contradiction in his Pulitzer Prize–winning book *The Denial of Death* when he called humans "gods with anuses": "Man is literally split in two: he has an awareness of his own splendid uniqueness in that he sticks out of nature with a towering majesty, and yet he goes back into the ground a few feet in order blindly and dumbly to rot and disappear forever. It is a terrifying dilemma to be in and to have to live with."[78]

8

OTHER THEORIES OF THE ORIGINS OF GODS

What a piece of work is man! how noble in reason! how infinite in faculty! in form and moving how express and admirable! in action how like an angel! in apprehension how like a god! the beauty of the world! the paragon of animals! And yet, to me, what is this quintessence of dust?

—WILLIAM SHAKESPEARE, *HAMLET*

Speculation regarding the gods has been ongoing for as long as gods have existed.

Indeed, the gods were prominent in *The Epic of Gilgamesh*, one of the earliest surviving works of literature. Such speculation has become increasingly prominent in the past two decades, especially in Europe, where only 21 percent of people "say that God plays an important role in their lives."[1]

This book has offered an evolutionary approach to the gods, an approach originally suggested by Charles Darwin. "A belief in all-pervading spiritual agencies seems to be universal," he noted, "and belief in spiritual agencies would easily pass into the belief in the existence of one or more gods." However, before this could happen, according to

Darwin, a "considerable advance in the reasoning powers of man" had to occur.[2]

Using neuroscience studies not, of course, available to Darwin, chapters 1 through 5 describe five major advances in "the reasoning powers of man." As hominin brains grew in size and developed increasingly strong connections among various brain areas, we acquired intelligence, an ability to think about ourselves, an ability to think about what others were thinking (theory of mind), and then an introspective ability to think about ourselves thinking about ourselves. Finally, about 40,000 years ago, we acquired an autobiographical memory, an ability to project ourselves backward and forward in time in a way not previously possible. We had become modern *Homo sapiens.*

The ability to project ourselves backward and forward in time profoundly affected the thinking of modern humans, since it allowed us to foresee our own deaths. Edward B. Tylor, a contemporary of Darwin, suggested that, in our quest to understand death, we identified the loss of the soul or spirit as the critical difference between life and death. Our new ability to integrate past, present, and future also enabled us to give meaning to our dreams in ways not previously possible. As Tylor noted, we experienced visits by deceased ancestors in our dreams and thus concluded that dead spirits continue to exist in an afterlife. This inevitably led to attempts to enlist and propitiate these spirits.

Once it was decreed that humans could continue to exist in another form after death, the seeds of the gods had been planted. As philosopher Sam Harris noted in *The End of Faith*: "A single proposition—*you will not die*—once believed, determines a response to life that would be otherwise unthinkable." Since deceased family members were believed to exist after death, it was logical to ask for their help and ancestor worship thus developed. This became increasingly elaborate and ritualized until ultimately a few very powerful ancestors broke through the celestial ceiling and were viewed as gods. This apparently occurred independently at several places in the world but could not be confirmed until written records became available.[3]

Our ability to integrate past, present, and future markedly improved our ability to plan and led directly to the agricultural revolution. As populations increased and became urbanized, driven by this revolution, secular authorities created rules and laws, and then aligned themselves with the gods to enforce them. Thus emerged the first religions, which invested the judicial, economic, and social needs of communities with the authority of the gods. As states and civilizations grew larger, so did religions. The influence of gods and religions was directly dependent on the influence of the civilizations to which they were tied, a pattern that continues to the present. Figure 8.1 represents these events schematically.

The brain evolution theory thus posits that gods, and later the formal religions that are tied to the gods, are products of the development of the human brain. Contemporary brain studies, as detailed in the foregoing chapters, confirm that our cognitive skills were acquired in

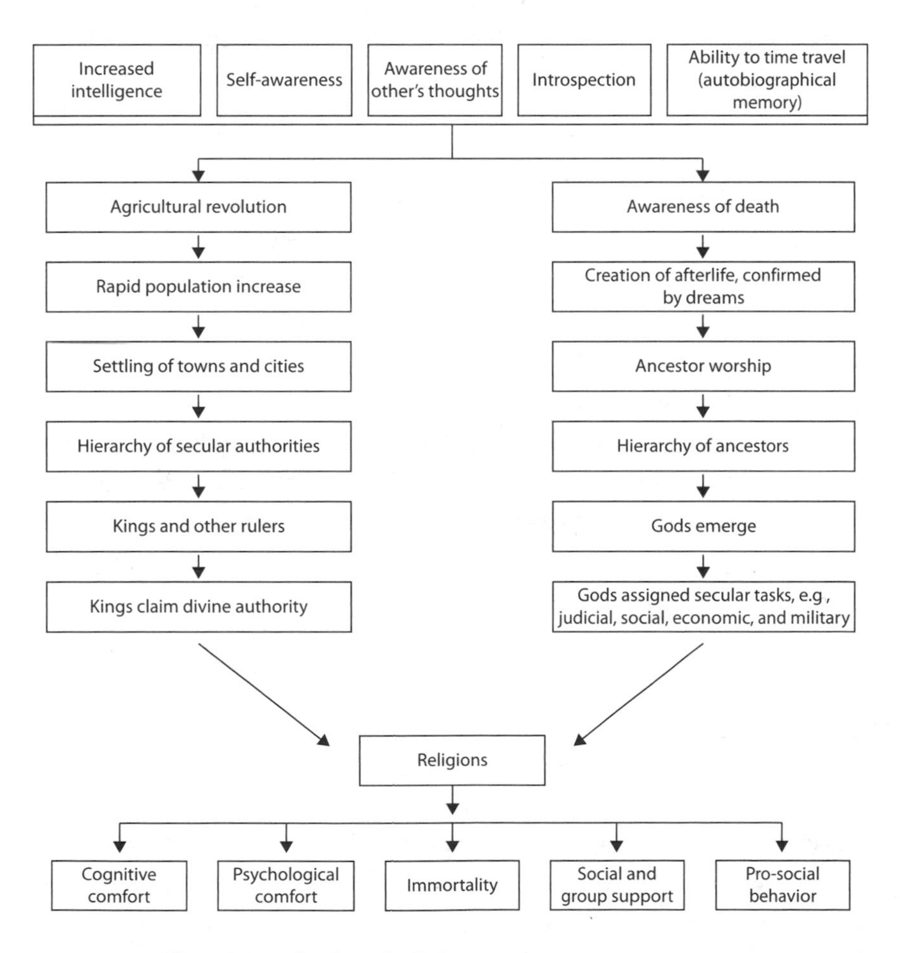

FIGURE 8.1 The origins of gods and religions.

the same order in which the brain areas associated with these skills evolved, thus providing anatomical support for this theory. Just as Darwin viewed a belief in "spiritual agencies" as having led to a belief in gods, so Tylor saw a belief in "the independent existence of the personal soul . . . in a Future Life" as having led to both gods and religions. As Tylor summarized this development:

> This great belief [in the continued existence of the soul] may be traced from its crude and primitive manifestations among savage races to its establishment in the heart of modern religion, where the faith in a future existence forms at once an inducement to goodness, a sustaining hope through suffering and across the fear of death, and an answer to the perplexed problem of the allotment of happiness and misery in the present world, by the expectation of another world to set this right.[4]

The brain evolution theory can thus explain both *why* the gods emerged and why they emerged *when* they did. Based on parallel evolution, it can also explain the independent emergence of gods at different places on earth. Finally, it can explain how the judicial, economic, and social needs of communities became joined with the spiritual needs of communities. The secular and the sacred developed together, each supportive of, and dependent on, the other.

Several other theories have been suggested to account for the origins of gods and religions. There is considerable overlap among these theories, and many scholars of the origin of gods and religions utilize more than one of the theories. At the risk of oversimplification, these theories are briefly summarized.

SOCIAL THEORIES

Social theories of the origins of gods and religions rely heavily on the work of Émile Durkheim, a nineteenth-century French thinker who is often cited as the founder of modern sociology. Durkheim believed that the origin of gods and religions lay not in spirits and dreams but rather in social structures and institutions. "The true nature of religion is to be found not on its surface, but underneath. . . . Religion's key value lies in the ceremonies through which it inspires and renews the allegiance of individuals to the group. These rituals then create, almost as an afterthought, the need for some sort of symbolism that takes the form of ideas about ancestral souls and gods." For Durkheim, "religion is something eminently social" and has its origin in the functions it serves. Indeed, gods are not essential for religion. For Durkheim, "religion and society are inseparable," and in his book *The Elementary Forms of the Religious Life*, published in 1912, he defined a religion as "a unified system of beliefs and practices relative to sacred things . . . which unite into one single moral community called a Church, all those who adhere to them."[5]

Durkheim strongly influenced most contemporary social theorists of religion. One example is *New York Times* reporter Nicholas Wade. In *The Faith Instinct*, Wade argued that "the evolutionary function of religion . . . is to bind people together and make them put the group's interests ahead of their own." Thus, said Wade, "groups with a stronger religious inclination would have been more united and at a considerable advantage compared with groups that were less cohesive." A religion, said Wade, "creates circles of trust whose members may support one another . . . [and] shapes members' social behavior, both toward one another (the in- group) and toward non-believers (the out-group)." A similar note was sounded by Barbara King, an anthropologist and primatologist at the College of William and Mary, in *Evolving God*. "Religion," she argued, "is built fundamentally upon belongingness," which she observed to be a powerful need in primates. Out of this need arose religion: "An earthly need for belongingness led to the human religious imagination and thus to the other-worldly realm of relating with God, gods, and spirits. From the building blocks we find in apelike ancestors emerged the soulful need to pray to gods, to praise God with hymns, to shake in terror before the power of invisible spirits."[6] David Sloan Wilson, an anthropologist at Binghamton University, is another prominent social theorist who emphasizes the social advantages of belonging to a religion. In his book *Darwin's Cathedral,* he cited as examples newly arrived immigrants who, by joining a church, can obtain help with such things as "buying a vehicle, finding housing, obtaining job referrals, baby-sitter referrals, Social Security information, . . . registering children for school, applying for citizenship, and dealing with the courts; the list of material benefits goes on and on." Social benefits also accrued to the temple followers of Enki and other gods in ancient Mesopotamia and thus have been inherent in the organization of religions from the beginning.[7]

Like other community organizations, religions do, of course, provide social benefits for their adherents and fulfill important social needs. This has been true since the emergence of the first recorded religions in Mesopotamia, as described in chapter 6. The question, however, is not whether religions fulfill social needs but whether this is the origin of gods and religions. Among some social theorists, gods are not prominent. For example, in *The Faith Instinct*, Nicholas Wade claims that "gods may not always be essential to an effective church." Under such theories, Thor and Zeus appear to have been stripped of their thunderbolts and instead appear as policemen or community organizers.[8]

PROSOCIAL BEHAVIOR THEORIES

Prosocial behavior theories of the origin of gods can be regarded as a special type of social theory. The core of such theories is that humans are being watched by the gods, the eye in the sky who sees and knows everything. Such theories stress the importance of gods and religions in enforcing social rules, morality, and group norms and suggest that "religion was invented to perpetuate a particular social order." Very useful in this regard is a belief in a god who sees and knows everything. The classic experiment that demonstrated the utility of such gods was an "honesty box" in a university coffee room, in which people were expected to deposit money for drinks they took. Over a 10-week period, the "honesty box" was decorated on alternating weeks with either a picture of flowers or a picture of a pair of eyes. During the weeks in which the eyes were used, almost three times more money was collected than during the weeks when flowers were used. The researchers concluded that "images of eyes motivate cooperative behavior because they induce a perception in participants of being watched." Interestingly, when a similar study was carried out on primary school children, the eyes had no effect on their behavior, presumably because the children had not yet acquired mature cognition.[9]

Three recent books focus on this theme that god is watching you. All three begin with the importance of theory of mind, as discussed in chapter 3, as the basis of religious belief. Once hominins acquired a theory of mind and understood that other people, as well as the gods, also had thoughts and emotions, they began an inevitable journey toward the formation of religions. In *The Belief Instinct*, Jesse Bering, a psychologist at Queen's University in Belfast, states it clearly: "God was born of theory of mind."[10]

This theme is carried forward by Ara Norenzayan, a psychologist at the University of British Columbia, in *Big Gods: How Religion Transformed Cooperation and Conflict*, and also by Dominic Johnson, a biologist at Oxford University, in *God Is Watching You: How the Fear of God Makes Us Human*. Both books claim that the concept of gods arose out of the theory of mind and our belief that the gods are watching us. Such a belief inclines us to cooperate with our fellow hominins; the more we cooperate, the more successful our group will be economically and socially, and the more of our genes will be passed on.[11]

Johnson suggests that the gods arose to satisfy our need to explain good and evil and the meaning of life: "Our brains are wired such that we *cannot help* but search for meaning in the randomness of life. It is human nature." To satisfy this need, we invented gods: "Human societies have invented gods not just once but thousands of times." By watching us and knowing what we are doing, the gods exert a positive force on us by promoting cooperative behavior: "The basic idea is that supernatural agents work like Big Brother looking over our shoulder, ever watchful, figures of both fear and awe that suppress our self-interest and make us more cooperative and productive." Such societies would have been more successful, according to Johnson, and thus evolutionarily more likely to pass along their genes to the next generation.[12]

Several other writers have stressed the importance of gods and religions in promoting moral and prosocial behavior. In *A History of God*, former Catholic nun Karen Armstrong argued that "without the idea of God there is no absolute meaning, truth or morality; ethics becomes simply a question of taste, a mood or a whim." Another contributor to such thinking is sociologist Robert Bellah. In his *Religion in Human Evolution*, he defined religion as "a system of beliefs and practices relative to the sacred that unite those who adhere to them in a moral community," and he described how shared communal activities such as play, ritual, and myth led to formal religions as societies became increasingly complex.[13]

Another variation on viewing religion as a mechanism for promoting social behavior was suggested by Boston University psychologist Patrick McNamara. In *The Neuroscience of Religious Experience*, he focused on the effects of religious belief and practices on individuals. McNamara distinguished the "current Self" from an "executive Self" and an "ideal Self." "Religion," he contended, "creates this executive Self by providing an ideal Self toward which the individual can strive and with which the individual can evaluate the current Self." Religious practices "are aimed at transforming the [current] Self into a higher, better Self. . . . Religion is interested in the Self because it seeks to transform the Self." McNamara described how religious practices affect the brain through a process he called "decentering." The ultimate goal of religion is thus to improve individual behavior and promote social cooperation, since "the executive Self is a social Self and is a master of social cooperation."[14]

It seems obvious that gods do play some role in promoting prosocial behavior, although the magnitude of that role may be debated. The question, however, is not whether the gods promote prosocial behavior and cooperation but rather whether that is the origin of the gods. Did the gods arise because of *Homo sapiens'* need for meaning and promotion of prosocial behavior, as prosocial theorists argue? Or did they arise in response to *Homo sapiens'* understanding of death and an afterlife, as this book argues, and then acquire a prosocial function later?

PSYCHOLOGICAL AND COMFORT THEORIES

The best-known psychological theory of the origins of gods and religions was put forth by psychoanalyst Sigmund Freud. According to Freud, our need to create gods as father figures arises from the unconscious need to resolve our Oedipal complex. This need arises in childhood when males wish to kill their fathers and marry their mothers, just as the Greek king Oedipus did. Thus, for Freud "religion arises *only* in response to deep emotional conflicts and weaknesses," and once people have resolved their unconscious conflicts thorough psychoanalysis, they will have no further need for religion.[15]

Although Freud's theories of religion fulfilling unconscious needs have been discredited, many contemporary theories emphasize the role of religions in fulfilling both conscious and unconscious needs for comfort. It is obviously comforting to look forward to going to heaven, being reborn, or entering some other form of afterlife rather than having to accept death as the end of one's existence. In most religions, the afterlife is portrayed as being very attractive. In the Mormon religion, for example, the Celestial Kingdom, which is the highest of the three levels of heaven, was described by the prophet Joseph Smith as having "beautiful streets . . . which had the appearance of being paved with gold." There, the faithful would dwell for eternity.

According to one Mormon scholar, "Each of us will look like ourselves, except that flaws, including potbellies, warts, deformities, and the like, will be gone, and a perfected physical figure—the very figure of our premortal spirit now clothed upon with immortal matter—will be ours." People will live in the Celestial Kingdom as families, and "those who died as babies and children will be raised to maturity by their exalted parents, in family chains that extend back to Adam and forward to forever."[16]

Most scholars of religion who regard the promise of an afterlife as an important factor in the origin of gods list it as one of several factors. Thus, Robert Hinde, a British zoologist at Cambridge University, wrote in *Why Gods Persist*: "Belief in a deity is related to a number of human propensities, especially understanding the causes of events, feeling in control of one's life, seeking security in adversity, coping with fear of death, the desire for relationships and other aspects of social life, and the search for a coherent meaning in life." Similarly, David Linden, a neuroscientist at Johns Hopkins University, acknowledged in *The Accidental Mind* that "religion provides comfort, particularly in allowing people to face their own mortality." He then gave equal weight to other advantages of religions, specifically that "religion allows for the upholding of a particular social order" and "religion gives answers to difficult questions."[17]

A neurochemical variation on the comfort theme was developed in *God's Brain*, by Lionel Tiger and Michael McGuire. They contended that "the experience of uncertainty and of facing the unknown" leads to chemical changes in our brains "that produce aversive physical and psychological states." As a reaction to such stress, our brains automatically regulate the brain's neurochemicals to produce what the authors call "brainsoothe." Religion is a major means of achieving "brainsoothe" by three mechanisms: the social aspects of religion produce pleasure by upregulating serotonin, dopamine, and norepinephrine; religious rituals relax the body; and religious beliefs simplify "the complexity of existence and social life." Thus, to understand religion, we need "to look at what religion does for the brain." The authors argue that "religion is to the brain what jogging is to the legs . . . a form of socioemotional and institutional exercise for the organ in our head."[18]

Some scholars have denied that a fear of death and a desire for an afterlife are especially important for the development of religion. For example, Pascal Boyer, an anthropologist at Washington University, argued that "the common shoot-from-the-hip explanation—people fear death, and religion makes them believe that it is not the end—is certainly insufficient because the human mind does not produce adequate comforting delusions against all situations of stress or fear." Boyer appears to categorize a fear of death as simply one of many human stresses and fears. Similarly, Stewart Guthrie, an anthropologist at Fordham University, claimed that some religions do not have an afterlife as part of their belief system. Therefore, he argued, "the lack of an afterlife, or of a happy one, found in many religions thus undermines two chief forms of the wish-fulfillment [comfort] theory: that belief is motivated by desire for immortality, and that it is motivated by a desire for posthumous retribution.[19]

To suggest that a fear of death and a desire for an afterlife are the most important factors in the emergence of gods is not to deny that other aspects of gods and religions are also comforting. As Hinde noted: "It is comforting to believe in a powerful entity who is on one's side, and who will intervene if appealed to." A belief in gods also suggests that someone is in control and that events have meaning. This is especially comforting in times of natural disasters such as earthquakes, floods, tornados, or typhoons. Gods are also comforting when people are faced with the deaths of loved ones or the random deaths of innocent children, or when good people are struck by lightning or a falling tree on their way to church. As Hinde noted: "Perhaps such issues can be encapsulated in the proposition that religious systems provide 'peace of mind,' a coherent view of the world, giving a semblance of order to a wide range of human experiences which might otherwise appear chaotic." As stated by Theodosius Dobzhansky: "Religion enables human beings to make peace with themselves and with the formidable and mysterious universe into which they are flung by some power greater than themselves."[20]

PATTERN-SEEKING THEORIES

Whereas psychological theories of the origin of gods and religions may provide psychological comfort, pattern-seeking theories may provide intellectual or cognitive comfort. Such theories have become prominent in recent years.

One of the first books to promote such theories was Stewart Guthrie's *Faces in the Clouds: A New Theory of Religion*, published in 1993. Guthrie, cited previously, argued that "religion may best be understood as systematic anthropomorphism: the attribution of human characteristics to nonhuman things or events." In fact, "anthropomorphism is the core of religious experience. . . . [It] pervades human thought and action, and . . . religion is its most systematic form." Not only do we naturally tend to see faces in clouds, he said, but we also attribute to gods natural phenomena, such as thunder and lightning. Guthrie claimed that anthropomorphizing is evolutionarily advantageous, "because the world is uncertain, ambiguous, and in need of interpretation." From an evolutionary point of view, he noted, "it is better for a hiker to mistake a boulder for a bear than to mistake a bear for a boulder."[21]

Over the past two decades, several pattern-seeking theorists have followed this line of reasoning. Psychologist and science writer Michael Shermer argued in *How We Believe* that "humans evolved to be skilled pattern-seeking creatures. . . . Humans evolved a Belief Engine whose function it is to seek patterns and find causal relationships. . . . Those who were best at finding patterns . . . left behind the most offspring." Pascal Boyer, cited previously, in *Religion Explained* described "religion in terms of cognitive processes that are common to all human brains, part and parcel of how a normal mind functions. . . . Faith and belief seem to be simple by-products of the way concepts and inferences are doing their work for religion in much the same way as 'for other domains.'" Similarly, Daniel Dennett, a philosopher at Tufts University, published *Breaking the Spell*, in which he argued that religious belief is the product of a human "hyperactive agent detection device." As he described it: "At the root of human belief in gods lies an instinct on a hair trigger: the disposition to attribute *agency*—beliefs and desires and other mental states—to anything complicated that moves."[22]

It is true that humans are pattern-seeking creatures, which is a direct result of the intelligence we developed over the past two million years. As discussed in chapter 4, however, could pattern-seeking, which is fundamentally an intellectual exercise, by itself have led to the origin of gods? The people who buried their kin at Sungir 28,000 years ago with elaborate grave goods and the people who built Göblecki Tepe 11,000 years ago were investing enormous resources into their labors and must have been moved by profoundly felt beliefs. Would pattern- seeking have been sufficient?

NEUROLOGICAL THEORIES

With the widespread availability of brain functional magnetic resonance imaging in recent years, there has been an outpouring of research efforts to identify brain areas associated with religious thinking. Such studies are often categorized as neurotheology and are well summarized in Patrick McNamara's *The Neuroscience of Religious Experiences.*[23]

The temporal lobe has been the focus of many studies, because individuals with temporal lobe epilepsy occasionally report religious experiences, such as seeing God, during their seizures. Vilayanur Ramachandran, a neuroscientist at the University of California at San Diego, reported that prior to such seizures one-quarter of such individuals "have deeply moving spiritual experiences, including a feeling of divine presence and the sense that they are in direct communication with God." Similarly, Michael Persinger, a psychologist at Canada's Laurentian University and author of *Neuropsychological Bases of God Beliefs*, contended that "the God Experience is a normal and more organized pattern of temporal lobe activity," a type of miniseizure "precipitated by subtle psychological factors such as personal stress, loss of a loved one, and the dilemma of anticipated death." Persinger believes that "a biological capacity for the God Experience was critical for the survival of the species. . . . The God Experience is a phenomenon that is associated with the construction of the temporal lobe. . . . If the temporal lobe had developed in some other way, the God Experience would not have occurred."[24]

The parietal lobe, especially the area adjacent to the superior temporal lobe (temporo- parietal junction), has also been the subject of neurotheology studies. When stimulated, this brain area may produce feelings of an out-of-body experience or "the feeling of a presence," often interpreted in a religious context. Persinger's studies have included the parietal area along with the temporal lobe. Similarly, Cosimo Urgesi et al. in Italy studied 88 individuals with brain tumors and reported that religious feelings of "self-transcendence" were associated with activity in the inferior parietal lobule. Matthew Alper, a philosopher and writer, labeled the temporal- parietal area "the God part of the brain" in his book of that title, and he whimsically suggested that someday it may be possible to surgically remove the "God part of the brain" by doing a "Godectomy."[25]

The hippocampus, the amygdala, and associated parts of the limbic system have also drawn attention in neurotheology studies. Rhawn Joseph, a psychologist who worked in the Palo Alto VA Hospital, theorized that the limbic system contains "God neurons" and "God neurotransmitters." A recent study by researchers at Duke University reported that atrophy of the hippocampus "was observed for participants reporting a life-changing religious experience." Patrick McNamara reported that "in hundreds of clinical cases and a handful of neuroimaging studies, it is a striking fact that the amygdala, large portions of the prefrontal lobes, and the anterior temporal cortex are repeatedly implicated in expression of religious experiences." McNamara labeled this a religion-related brain circuit.[26]

Consistent with McNamara's findings, the frontal lobe has also figured prominently in studies of neurotheology. For example, Andrew Newberg and Eugene d'Aquili at the University of Pennsylvania studied brain areas activated when Franciscan nuns and Buddhist monks were meditating and reported "greater activity in the frontal lobes, and in particular in the prefrontal cortex." At the same time, the nuns and monks showed reduced activity in their parietal lobes and described themselves "as entering a state of timelessness and spacelessness." Other researchers have related religiosity to specific parts of the frontal lobes, including the orbital frontal cortex and the anterior cingulate, or to a combination of frontal and parietal areas.[27]

Still other researchers have linked religious ideation to other brain areas, such as the thalamus and caudate, as well as to specific neurochemical systems, such as dopamine and serotonin. What seems clear at this point is that there is no single "god center" in the brain. Rather, religious experiences are mediated by an extensive brain network similar to the network, described in the foregoing chapters, that mediates awareness of self, awareness of others, introspection, and autobiographical memory—in other words, the brain network that makes us uniquely human. McNamara similarly noted that "there is considerable anatomical overlap between the brain sites implicated in religious experience and the brain sites implicated in the sense of Self and self-consciousness." It is also clear that the areas of the brain that are activated by religious experiences depend on the specific type of that experience. For example, meditation will activate frontal areas, whereas experiences involving intense emotions will activate the amygdala. Similarly, a study in which some subjects were asked to experience "an intimate relationship with God" while other subjects were asked to experience "fear of God's anger" also activated different brain areas.[28]

GENETIC THEORIES

Studies of twins have suggested that there is a genetic component to religiosity. A study of teenage twins, both identical and nonidentical, "found the genetic contribution toward variation in religiosity to be around 20 percent." Another widely publicized study of identical and nonidentical twins reared apart assessed religiosity in a variety of ways (for example, religious beliefs, interest in religious occupations) and reported a "50 percent genetic influence on religiosity." The authors of that study cautioned, however, that the genetic influence "might operate through personality traits such as traditionalism"; in other words, individuals who inherit similar personality traits might be more drawn to religious ideas. In such cases, the genetic effect would be on the personality traits, not on religiosity as such.[29]

A few researchers have even suggested that "our universal spiritual/religious proclivities represent . . . a genetically inherited trait . . . what we could call 'spiritual' genes." If true, then "human beings are genetically predisposed or 'hard-wired' to believe in the concepts of spiritual reality, a God or gods, a soul, and an afterlife."[30]

The most ambitious attempt to identify such a gene was carried out by geneticist Dean Hamer, who in 2004 published *The God Gene* and was featured in a cover story in *Time* magazine. Hamer used a "self-transcendence scale of spirituality," which included questions such as connectedness to nature and interest in extrasensory perceptions, as his measure of religiosity. He then identified a gene that accounted for only 1 percent of the variance in the test scores of his subjects and designated it a "spiritual allele," or "God gene." The designated gene affects dopamine, serotonin, and other brain chemicals that, when released, according to Hamer, "bring a profound sense of joy, fulfillment, and peace." Hamer's work has been widely criticized for his choice of measures of religiosity, for his statistically weak findings, and for his designation of a single gene as a "God gene" when it is known that most human traits are the product of hundreds of genes, if they are genetic in origin at all.[31]

Another widely cited, if quixotic, attempt to establish a genetic basis for religiosity was a book by psychologist Julian Jaynes, *The Origin of Consciousness and the Breakdown of the Bicameral Mind*, published in 1976. Jaynes argued that until about 3,000 years ago, the two halves of the brain, the "bicameral mind," operated independently. At that time, a genetic change took place, causing the two halves of the brain to become integrated. This in turn produced auditory hallucinations that humans interpreted as the voices of gods, and this gave rise to religions. As Jaynes summarized it: "The neurological structure responsible for these hallucinations is neurologically bound to substrates for religious feelings, and this is because the source of religion and of gods themselves is in the bicameral mind." Jaynes's thesis is at odds with almost everything known about the evolution of the human brain.[32]

ARE GODS THE PRODUCTS, OR BY-PRODUCTS, OF EVOLUTION?

A final question regarding theories of the origin of gods is whether the emergence of the gods represents an adaptation of evolution and was evolutionarily advantageous, or whether the emergence of gods was merely a by-product of evolution, "a vestigial artifact of a primitive mind," in the words of one writer. Debate on this issue has been ongoing and spirited, with the majority of writers favoring an adaptationist position.[33]

The most common argument put forth by the adaptationists is that gods improve the survival of the group. According to this theory, "ancestral societies with culturally widespread God concepts would have outcompeted societies without such concepts, given the cooperative advantage of believing groups." This argument assumes that groups sharing gods are more willing to share resources, more willing to defend the group against outside threats, and more cooperative in general. As summarized by Nicholas Wade: "Other things being equal, groups with a stronger religious inclination would have been more united and at a considerable advantage compared with groups that were less cohesive. People in the more successful group would have left more surviving children, and genes favoring an instinct for religious behavior would have become commoner each generation until they had swept through the entire human population." Although this is a reasonable hypothesis, I am not aware of any supporting data. In addition, some geneticists have questioned the validity of group selection, saying that evolutionary theory applies only to individuals.[34]

At the individual level, there are also said to be evolutionary advantages to believing in gods. Dean Hamer argued that the "God genes" are advantageous in providing "human beings with an innate sense of optimism. . . . Optimism is the will to keep on living and procreating, despite the fact that death is ultimately inevitable." Matthew Alper similarly wrote that "those individuals whose brains possessed some genetic mutation that could withstand the overwhelming anxiety induced by our awareness of death were more likely to survive." And Patrick McNamara contended that the religion-inspired "unified Self" may be "more effective in pursuing behavioral goals . . . in evading predators . . . in war and combat," and more cooperative.[35]

Another argument in favor of gods being evolutionarily advantageous is that they are good for your physical and mental health. Numerous studies have reported that individuals who attend church regularly have lower rates of hypertension, heart disease, emphysema, cirrhosis, anxiety, depression, and suicide. However, individuals who attend church regularly may also be less likely to smoke and drink heavily, which may account for many of these differences. In addition, most such studies have been carried out on older adults; such differences would only be evolutionarily advantageous if they were operational for individuals in their childbearing years.[36]

On the other side of the argument, fewer scholars of religion have portrayed gods as by-products of evolution. The brain evolution theory outlined in this book proposes that gods are a by-product of our acquisition of autobiographical memory and suggests that religions followed the emergence of gods as the population increased and societies became organized. Other writers who have defended a by-product position include Pascal Boyer in *Religion Explained*, cited previously, who argued that gods and religions are by-products of the human tendency to seek patterns. Scott Atran, an anthropologist at the Centre National de la Recherche Scientifique in Paris, claimed in his book *In Gods We Trust* that "religion has no evolutionary function per se." And Richard Dawkins, a biologist at Oxford University, argued in *The God Delusion* that religion "doesn't have a direct survival value of its own, but is a by-product of something else that does."[37]

As by-products of evolution, gods would usually be assumed to be neutral and to have had no effect on evolution as such. This may or may not be true, since it is also possible that gods may be ultimately evolutionarily disadvantageous. Possible scenarios for this are "god contests," in which wars are fought to determine whose god is the correct god. Such wars were fought between city-states in ancient Mesopotamia, as described in chapter 6, and apparently contributed to the demise of the world's first civilization. An Old Testament god contest familiar to many is the battle between the followers of Baal, the Canaanite fertility god, and the followers of Jehovah, the Israelite protector God. Elijah, a prophet of Jehovah, prevailed and then had the 450 followers of Baal put to death.[38]

The history of modern *Homo sapiens* up to the present is littered with god contests. In *The Great Big Book of Horrible Things*, a list of history's 100 worst manmade atrocities, 25 of the 100 were god contests. Such contests become especially dangerous when combined with apocalyptic beliefs about the end of the world as being glorious and weapons of mass destruction capable of terminating the existence of modern *Homo sapiens*. In his *Letter to a Christian Nation*, Sam Harris invoked this apparition:

> Imagine the consequences if any significant component of the U.S. government actually believed that the world was about to end and that its ending would be glorious. The fact that nearly half of the American population apparently believes this, purely on the basis of religious dogma, should be considered a moral and intellectual emergency. . . . That religion may have served some necessary function for us in the past does not preclude the possibility that it is now the greatest impediment to our building a global civilization.

Under such scenarios, the deistic by-products of evolution could terminate human existence with one final nuclear chorus of *Dies Irae*.[39]

Humans need gods. As Fyodor Dostoevsky phrased it: "Man needs the unfathomable and infinite just as much as he needs the small planet which he inhabits." Since the human need for gods is an integral part of the brain networks that make us uniquely human, and since formal religions are deeply socially integrated into our cultures, neither gods nor religions are likely to simply disappear anytime soon, even if they are no longer needed. In America alone, there are over 1,500 different religious denominations, ranging from the Agasha Temple of Wisdom to Zygon International; most of these are small, but 25 of them have at least one million adherents. At many other places in the world, gods and religions also continue to play very important roles in the lives of the people. The Jotabeche Methodist Pentecostal Church in Santiago, Chile, seats 18,000 people. The Yoide Full Gospel Church in Seoul, South Korea, seats 12,000 in the main church and 20,000 in overflow chapels and has seven Sunday services. As British anthropologist Sir James Frazer noted: "It seems probable that the great majority of our species will continue to acquiesce in a belief so flattering to human vanity and so comforting to human sorrow." Frazer continued: "It cannot be denied that the champions of eternal life have entrenched themselves in a strong, if not impregnable, position; for if it is impossible to prove the immortality of the soul, it is, in the present state of our knowledge, equally impossible to disprove it."[40]

Thus, gods and their accompanying religions will probably continue to be born and to die. Examples of religions that have come into being within the last two centuries and that already have several million adherents are Ahmadiyya in Pakistan and Mormonism in the United States. Ahmadiyya was founded by Mirza Ghulam Ahmad, who claimed to be the promised Messiah and Mahdi awaited by Muslims. Ahmadis teach that Jesus was an earlier prophet who was crucified but survived his time on the cross and died later of old age in Kashmir, where he was seeking the Lost Tribes of Israel. Ahmadiyya has been influential among African American Muslims. Mormonism, officially known as the Church of Jesus Christ of Latter-Day Saints, was founded by Joseph Smith, who claimed to have found buried in the ground, where an angel directed him to dig, golden plates that were the records of an ancient religion. This religion had been founded by Israelites who had come to America 2,600 years earlier. Mormons teach that, following his crucifixion and rising from the dead, Jesus came to America and designated it as the new Promised Land. The Mormon religion thus claims to be a continuation of this ancient religion.[41]

Just as new gods and religions will continue to be born, so others will continue to die. Many old gods, such as Anu, Ra, Zeus, and Jupiter, grace the world's art museums, admired but not revered, viewed as artistic creations rather than divine creations. Such museums should properly be regarded as shrines to dead gods. Other old gods have been adopted by New Age or other contemporary religions. At Avebury and Stonehenge, the solstice is celebrated with hymns to druid gods by the Secular

Order of Druids, despite the fact that druids did not exist until 2,000 years after these monuments were built. Similarly, Çatalhöyük in Turkey has become a pilgrimage site for goddess worshipers.[42]

Viewing monuments originally built to honor the old gods also provides a needed historical perspective. In Gloucestershire, England, a large burial mound, perhaps 5,000 years old, is said to be "one of the best places in the area for flying kites and model aircraft." At a nearby mound, in which 24 skeletons were found, "barely a summer Sunday afternoon passes without picnickers." In Newark, Ohio, the extraordinary Great Circle Earthworks, sacred burial mounds created by the Hopewell people 2,000 years ago, have been incorporated into the Moundbuilders Country Club's 18-hole golf course. Some burial mounds serve as tees, while others act as bunkers surrounding the holes. The ninth tee, for example, is on top of an eight-foot mound, and the 219-yard, par 3 hole follows the passage that originally divided the ancient octagon from the great circle, a likely route for stately religious processions. Included on the country club's website is the following note: "Won't archeologists 2000 years from now be puzzled as they study the mounds and find all those lost golf balls?"[43]

Most of the old gods and religious monuments, however, have been lost to human history. They stand, like Ozymandias, as "two vast and trunkless legs of stone" in the desert:

> Look on my Works, ye Mighty, and despair!
> Nothing beside remains. Round the decay
> Of the colossal Wreck, boundless and bare
> The lone and level sands stretch far away.[44]

APPENDIX A

THE EVOLUTION OF THE BRAIN

In order to understand the evolution of the human brain, it is useful to consider exactly what it is in the brain that is undergoing evolution. Most obvious, of course, is the size of the brain, both its absolute size and its relative size. In an absolute sense, whales and elephants have much larger brains that humans do, but relative to the size of their bodies they have much smaller brains than humans. More important than the overall size of the brain is the size of specific brain areas. In humans, for example, the frontal pole (BA 10), an area associated with many cognitive functions discussed in this book, is said to be "twice the size of what we would expect" based on comparisons with other primates. Brain size is discussed at length in chapter 2.[1]

During human evolution, the neurons, glial cells, and connecting fibers all underwent changes. The neurons increased in number and became more closely packed, so that humans have 25,000 to 30,000 neurons per cubic millimeter of brain cortex. By comparison, whales and elephants have only 6,000 to 7,000 neurons per cubic millimeter of brain cortex. Glial cells, which are 10 times more numerous than neurons, also underwent evolutionary changes. Especially important are the glial cells that make myelin coating for the connecting fibers, because it is the myelin coating, in combination with the diameter of the nerve fiber, that speeds up the transmission of information on the connecting fibers.

Humans have thick myelin coverings on the connecting fibers, whereas whales and elephants have very thin myelin coverings; this is a major reason why the transmission of information in human brains is up to five times faster than in whales and elephants.[2]

The relative importance of glial cells and connecting fibers in human brain evolution was illustrated by a comparative study of gray matter (neurons) and white matter (glial cells and connecting fibers) in the prefrontal brain region in chimpanzees and humans. It was expected that the major difference between chimpanzees and humans would be in the gray matter.

However, humans had only 2 percent more gray matter than chimpanzees but 31 percent more white matter. In a related study, it was reported that brain connections develop much faster in human infants than they do in chimpanzee infants. Such studies suggest that "humans may have an enhanced ability to integrate information across modalities in comparison to other primates" and that it is our white matter connecting tracts more so than our gray matter neurons that make us uniquely human.[3]

The ascertainment of which brain areas developed early in the evolution of hominins and which areas developed more recently is important. Three measures are most commonly used.

The most widely used method is the development of myelin around the connecting nerve fibers. It is the myelin that speeds up the transmission of information on the nerve fibers. The process of myelination begins while the developing brain is still in utero and continues after birth, through adolescence, and into the person's twenties. The order in which the myelination of the nerve fibers takes place is assumed to reflect the order in which these brain areas evolved. As stated by Orthello Langworthy, an embryologist at Johns Hopkins University who did detailed studies of myelination, "The pathways in the nervous system become myelinated in the order in which they were developed phylogenetically."[4]

The definitive study of myelination was carried out in the 1890s by a German researcher, Paul Emil Flechsig, who studied the brains of deceased infants, ranking 45 brain areas by their degree of myelination. Nine areas, or 20 percent of the total, were found to be the least myelinated in the infant brains; these areas were assumed to have evolved most recently and were referred to by Flechsig as "terminal zones." Even though the "terminal zones" include only 20 percent of the brain areas listed by Flechsig, it is significant that they include most of the areas discussed in this book as being associated with the cognitive abilities that make us uniquely human. The exceptions to this rule are a few evolutionarily older brain areas, such as the hippocampus and cerebellum, that were altered later in evolution to accommodate more recently derived functions, as described in chapter 6.[5]

Postmortem human brains offer a second measure to assess which brain regions evolved most recently. This is the relative degree of infolding, or gyrification, as it is also called, of various brain regions. As primate brains evolved, they developed progressively more infoldings; such infoldings allowed the brain to increase its surface area without the brain having to grow larger in size. Thus, humans have 49 percent more infoldings than rhesus monkeys and 17 percent more infoldings than chimpanzees. Within human brains, different regions have different degrees of infolding. This has been studied by German anatomist Karl Zilles and his colleagues, who ranked the regions in human brains using a gyrification index. The two most infolded brain regions, and thus the most recently evolved, are the prefrontal cortex and parietal lobe, both of which include areas critical for the cognitive abilities that make us uniquely human. Zilles et al. concluded that "a higher degree of infolding . . . is interpreted as an indicator of a progressive evolution of this cortical region in humans." Related to this is a third measure of maturation, which is the degree to which the brain infoldings are similar from one person to another. Almost a century ago, it was noted that individual anatomical variability "suggests at once, if it exceeds rather narrow limits, that the organ or structure concerned has not yet reached its full development"; in other words, a greater variation in the pattern of infoldings indicates that that brain area has evolved more recently.

The inferior parietal area, which is very important for our discussion, is well known among neuroanatomists for "the bewildering variety of the sulcal [groove] patterns in this region," an indication of its recent origin. Other researchers have also commented on the "very considerable variation" in the pattern of infoldings in both frontal and parietal areas in human brains.[6]

APPENDIX B

DREAMS AS PROOF OF THE EXISTENCE OF A SPIRIT WORLD AND LAND OF THE DEAD

The Human Relations Area Files (HRAF) is a nonprofit organization at Yale University. It was founded in 1949 to bring together in one place nineteenth- and twentieth-century ethnographic accounts of cultures from around the world. Since 1994 these accounts have been available online (http://ehrafworldcultures.yale.edu). As of June 2016, HRAF files included 295 cultures, which are categorized by subsistence type. Seventy-one of the 295 cultures are hunter-gatherers, defined as depending almost entirely or largely on hunting, fishing, and gathering for subsistence. The following are some accounts of dreams in these hunter-gatherer societies from the HRAF files.

CREEK INDIANS OF SOUTHEASTERN STATES

The body is buried with personal possessions and food offerings for the journey and monthly offerings are left at the grave for the first year. The spirits of the dead are believed to appear in dreams to advise the living. (Richard A. Sattler, *Culture Summary: Creek* [New Haven, CT: Human Relations Area Files, 2009], http://ehrafworldcultures.yale.edu/document?id=nn11-000.)

COMANCHE INDIANS OF THE GREAT PLAINS

Religious patterns included a belief in a vaguely defined Great Spirit who was the fountainhead of all power but who did not interfere in human affairs. This power could be obtained by men through dreams in which a supernatural patron or guardian spirit endowed the petitioner with a certain amount of power and various songs and procedures needed by the recipients to manipulate his Medicine. (David E. Jones, *Sanapia, Comanche Medicine Woman* [New York: Holt, Rinehart and Winston, 1972], http://ehrafworldcultures.yale.edu/document?id=n006-031.)

UTE INDIANS OF UTAH AND COLORADO

When you dream about your dead relatives *n'saka'*, they are just trying to tell you something that you should do. In most instances dreams are attributed to personalized spirit beings, yet sometimes they emanate from the impersonal wellsprings of power itself. (Joseph G. Jorgensen, *Ethnohistory and Acculturation of the Northern Ute* [Ann Arbor, MI: University Microfilms, 1980], http://ehrafworldcultures.yale.edu/document?id=nt19-019.)

EASTERN APACHE INDIANS OF THE SOUTHWEST

Most often it is in dreams that the dead are distinctly seen:

> Ghosts appear also in dreams, in sleep. That is the worst form, I guess. You really see them in a dream. I get like that. The door opens and they get closer and closer. I want to get up and fight, but I can't move. I can just say, "Ah!" The Chiricahua say this is ghost sickness. It can make a person very ill. It's a sign of trouble with evil ghosts if you do that too much, and you have to go to a shaman about it.

(Morris Edward Opler, *An Apache Life-Way: The Economic, Social, and Religious Institutions of the Chiricahua Indians* [Chicago: University of Chicago Press, 1941], http://ehrafworldcultures.yale.edu/document?id=nt08-001.)

POMO INDIANS OF CALIFORNIA

The reason for burning the possessions of a dead person was not that the dead might use the objects in the ghost world, but because the possessions would be rendered impure by ghostly visitations. Ghosts returned in the form of dreams and haunted their cherished possessions. (Edwin Meyer Loeb, *Pomo Folkways*, Publications in American Archaeology and Ethnology [Berkeley: University of California Press, 1926], http://ehrafworldcultures.yale.edu/document?id=ns18-003.)

YORUK INDIANS OF CALIFORNIA

The prophets [of the Yoruk] visited the dead in dreams and carried messages from them—once even that they would appear the next day. (A. L. [Alfred Louis] Kroeber, *Handbook of the Indians of California*, bulletin [Washington: Government Printing Office, 1925], http://ehrafworldcultures.yale.edu/document?id=ns31-009.)

TLINGIT INDIANS OF ALASKA

In any case, it denotes the entity which "lives" after death, which returns to visit the living in dreams, and which becomes reincarnated. It would appear, therefore, to apply to the essential "self" of the person. (Frederica De Laguna, *Under Mount Saint Elias: The History and Culture of the Yakutat Tlingit*, Smithsonian Contributions to Anthropology

[Washington, DC: Smithsonian Institution Press, 1972; for sale by the Supt. of Docs., U.S. Govt. Print. Off.], http://ehrafworldcultures.yale.edu/document?id=na12–020.)

OJIBWA INDIANS OF CANADA

A further link between persons of this category and human beings of the grandparent class is the fact that, collectively, other-than-human persons were referred to as "our grandfathers." Besides this, the Ojibwa believed that they came into direct personal contact with other-than-human persons in their dreams. (A. Irving [Alfred Irving] Hallowell, "Northern Ojibwa Ecological Adaptation and Social Organization," in *Contributions to Anthropology: Selected Papers of A. Irving Hallowell* [Chicago: University of Chicago Press, 1976], http://ehrafworldcultures.yale.edu/document?id=ng06–067.)

STONEY (NAKODA) INDIANS OF CANADA

Even if no specific vision was granted the seeker, the Great Spirit's presence was never doubted. In times past He appeared and revealed Himself in various ways. He appeared in dreams, visions, and sometimes He spoke to us through the wild animals, the birds, the winds, the thunder, or the changing seasons. (John Snow, *These Mountains Are Our Sacred Places: The Story of the Stoney Indians* [Toronto, Ontario, Canada: Samuel-Stevens, 1977], http://ehrafworldcultures.yale.edu/document?id=nf12–027.)

CREE INDIANS OF CANADA

Manitous or spirits could inhabit all living things, as well as objects or forces (such as wind and thunder); and many of these were considered animate. Manitous appeared in dreams and gave special power or protection to the individual. Some men obtained great powers from the manitous; in curing they called upon manitous for help. (James G. E. Smith, "Western Woods Cree," in *Handbook of North American Indians: Subarctic*, ed. June Helm [Washington, DC: Smithsonian Institution, 1981; for sale by the Supt. of Docs., U.S. G.P.O.], http://ehrafworldcultures.yale.edu/document?id=ng08-002.)

BELLA COOLA INDIANS OF WESTERN CANADA

By dreams a man learns the fortune of his supernatural representative, and from them he can judge what is in store for him during the coming year and receive intimations concerning birth and death, secret society matters and, in fact, every phase of human activity, since all are decided at the meeting of Äłquntäm and his associates. Dreams are considered to be especially important at this season of the year and information obtained from them may, perhaps, be the origin of the present firm conviction that some portion of the body actually ascends. (T. F. [Thomas Forsyth] McIlwraith, *Bella Coola Indians: Volume One* [Toronto: University of Toronto Press, 1948], http://ehrafworldcultures.yale.edu/document?id=ne06-001.)

NOOTKAN INDIANS OF WESTERN CANADA

People frequently see the dead in dreams and this is regarded as good evidence for the nature of the life of the dead. (Elizabeth Colson, *The Makah Indians: A Study of an Indian Tribe in Modern American Society* [Minneapolis: University of Minnesota Press, 1953], http://ehrafworldcultures.yale.edu/document?id=ne11-002.)

CHIPEWYAN INDIANS OF NORTHERN CANADA

The Chipewyan were—and generally are—animists. Animals, spirits, and other animate beings existed in the realm of INKOZE simultaneously with their physical existence. Humans were part of the realm of INKOZE until birth separated them from that larger domain for the duration of their physical existence. Knowledge of INKOZE came to humans in dreams and visions given them by animals or other spirits. . . . The dead retain a recognizable identity as they may visit the living in dreams or visions. The Christian concept of the soul has been added onto traditional beliefs about the spiritual construction of the person without displacing them. (Henry S. Sharp and John Beierle, *Culture Summary: Chipewyans* [New Haven, CT: Human Relations Area Files, 2001], http://ehrafworldcultures.yale.edu/document?id=nd07–000.)

INUIT OF THE CANADIAN ARCTIC

Because Kirluayok warned us that your medicine would kill us. He said the Spirits of the Sea and Land told him in his dreams that we must avoid touching or receiving anything from you, or we shall all die. (Raymond De Coccola, Paul King, and James Houston, *Incredible Eskimo: Life Among the Barren Land Eskimo* [Surrey, BC: Hancock House, 1986], http://ehrafworldcultures.yale.edu/document?id=nd08–035.)

BARAMA RIVER CARIBS OF BRITISH GUIANA

If a man dreams of a dead person, he believes that he is actually seeing the ghost of the dead walking by at that moment. (John Gillin, *The Barama River Caribs of British Guiana*, Papers of the Peabody Museum of American Archaeology and Ethnology [Cambridge, MA: Museum, 1936], http://ehrafworldcultures.yale.edu/document?id=sr09–001.)

MATACO INDIANS OF BOLIVIA

Honhat is also the home of both natural and supernatural forces. Man cannot go there, except in dreams and in ecstasy. It is the place of the dead and the illnesses and therefore regarded as more genuinely evil. (Jan-åke Alvarsson, *The Mataco of the Gran Chaco: An Ethnographic Account of Change and Continuity in Mataco Socio-Economic Organization*, Acta Universitatis Upsaliensis, Uppsala Studies in Cultural Anthropology [Uppsala, Sweden: Academiae Upsaliensis, 1988; distributed by Almqvist and Wiskell International], http://ehrafworldcultures.yale.edu/document?id=si07-009.)

Very often in dreams one sees dead relatives. The soul has gone to the underworld and paid them a visit. Sometimes the dead return by night to the houses of the living who then dream of the departed. (Alfred Métraux, *Myths and Tales of the Matako Indians (The Gran Chaco, Argentina)*, Ethnological Studies [Gothenburg, Sweden: Walter Kaudern, 1939], http://ehrafworldcultures.yale.edu/document?id=si07-003.)

CANELA INDIANS OF BRAZIL

Ghosts visit some youths who are seriously trying to become shamans but not others. They may visit a person unexpectedly when he is sick to make him a shaman. They travel in the other world in dreams, or in their belief, and often go to the land of the dead to bring a wandering soul back to its body, saving its life. (William H. [William Henry] Crocker and John Beierle, *Culture Summary: Canela* [New Haven, CT: Human Relations Area Files, 2012], http://ehrafworldcultures.yale.edu/document?id=so08-000.)

MBUTI PYGMIES OF CENTRAL AFRICA

Hallucinations and dreams are primarily the result of accidentally slipping from the one world into the other. . . . Dreams, then, do not convey authority as being divine portents but, being real experience of a mirror world, they are, like all experiences, to be learned from. (Colin M. Turnbull, *Wayward Servants: The Two Worlds of the African Pygmies* [Garden City, NY: Natural History Press, 1965], http://ehrafworldcultures.yale.edu/document?id=fo04-002.)

KALAHARI SAN OF SOUTHERN AFRICA

These states, whether dreams, trances, or day-time confrontation with the spirits, are regarded as reliable channels for the transfer of new meaning from the other world into this one. (Megan Biesele, *Women Like Meat: The Folklore and Foraging Ideology of the Kalahari Ju/'Hoan* [Johannesburg, South Africa: Witwatersrand University Press; Bloomington: Indiana University Press, 1993], http://ehrafworldcultures.yale.edu/document?id=fx10-067.)

VEDDAHS OF CEYLON

Every near relative becomes a spirit after death, who watches over the welfare of those who are left behind. These, which include their ancestors and their children, they term their "néhya yakoon," kindred spirits. They describe them as "ever watchful, coming to them in sickness, visiting them in dreams, giving them flesh when hunting." (John Bailey, "An Account of the Wild Tribes of the Veddahs of Ceylon: Their Habits, Customs, and Superstitions," *Transactions* 2 [1863]: 278–320, http://ehrafworldcultures.yale.edu/document?id=ax05-002.)

ANDAMAN ISLANDERS

In his dreams he can communicate with the spirits of the dead. (A. R. [Alfred Reginald] Radcliffe-Brown, *The Andaman Islanders: A Study in Social Anthropology* [Cambridge: Cambridge University Press, 1922], http://ehrafworldcultures.yale.edu/document?id=az02–001.)

MANUS ISLANDERS

At one stage, a few of the men even planned to go to Baluan, but then something unexpected happened. Suddenly some of the islanders became stricken by an attack of guria, or violent shaking. The phenomena of guria was accompanied by dreams through which the villagers received messages, that the ancestors were to return with large quantities of cargo, and to help them build a new society. (Berit Gustafsson, *Houses and Ancestors: Continuities and Discontinuities in Leadership Among the Manus* [Göteborg: IASSA, 1992], http://ehrafworldcultures.yale.edu/document?id=om06–010.)

BATEK OF MALAYSIA

Generally speaking, the Lebir Batek Dè' emphasize the similarities between humans and the hala' 'asal rather than the differences. They see human beings as becoming virtually identical to the hala' 'asal after death, when the shadow-souls of the dead acquire young bodies and water life-souls. The rejuvenated dead also live with the hala' on top of the firmament. Like the hala', they spend much of their time singing and decorating themselves with flowers, and they come to earth from time to time where they are seen by the living in dreams. (Kirk M. Endicott, *Batek Negrito Religion: The World-View and Rituals of a Hunting and*

Gathering People of Peninsular Malaysia [Oxford: Clarendon; Oxford University Press, 1979], http://ehrafworldcultures.yale.edu/document?id=an07-004.)

KORYAK OF EASTERN RUSSIA

I would like to point out that Koryak vampires are known as a kind of shaman. They interact with autochthonous spirits to cure or divine, travel to the land of the dead in trance or dreams. (Alexander D. King, "Soul Suckers: Vampiric Shamans in Northern Kamchatka, Russia," *Anthropology of Consciousness* 10, no. 4 [1999]: 57–68, http://ehrafworldcultures.yale.edu/document?id=ry04-032.)

AINU OF SAKHALIN ISLAND IN RUSSIA

When people dream, their soul frees itself from its sleeping owner's body and travels to places distant in time and space. This is why in our dreams we visit places where we have never been. By the same token, a deceased person may appear in our dreams, because the soul can travel from the world of the dead to visit us during our dreams. (Emiko Ohnuki-Tierney, *Illness and Healing Among the Sakhalin Ainu: A Symbolic Interpretation* [Cambridge: Cambridge University Press, 1981], http://ehrafworldcultures.yale.edu/document?id=ab06-013.)

NOTES

PREFACE

1. Carl Zimmer, *Soul Made Flesh: The Discovery of the Brain—and How It Changed the World* (New York: Free, 2005), 174. Zimmer's account of Willis's work is superb, including the fact that Christopher Wren was doing Willis's drawings for him.
2. The Upper Paleolithic period is commonly divided into four subperiods based on the cultural artifacts that have been found dating to that time. These are the Aurignacian (45,000–28,000 years ago), Gravettian (28,000–21,000 years ago), Solutrean (21,000–18,000 years ago), and Magdalenian (18,000–11,000 years ago). Some authors also refer to the period from 14,000 to 12,000 years ago as the Epipaleolithic period. The Upper Paleolithic period was then followed by the Neolithic period, beginning about 11,000 years ago.
3. William James, *The Varieties of Religious Experience* (New York: Random House, 1929), 31–34.

INTRODUCTION

1. Carl Jung, *The Integration of the Personality* (London: Routledge and Kegan Paul, 1950), 72; Patrick McNamara, *The Neuroscience of Religious Experience* (New York: Cambridge University Press, 2009), ix.
2. Pew Forum on Religion and Public Life, *"Nones" on the Rise: One-in-Five Adults Have No Religious Affiliation* (Washington, DC: Pew Forum on Religion and Public Life, 2012), www.pewforum.org/unaffiliated/nones-on-the-rise.aspx; Harris Poll #90, The Religious and Other Beliefs of Americans, 2005, Harris Interactive, December 14, 2005, www.harrisinteractive.com/harris_poll/index.asp?PID=618; M. Lilla, "The

Politics of God," *New York Times Magazine*, August 19, 2007, 28–35, 50, 54–55, quoting Rousseau; Francis Collins, *The Language of God: A Scientist Presents Evidence for Belief* (New York: Free, 2006), 38, 149; S. Begley, "In Our Messy, Reptilian Brains," *Newsweek*, April 9, 2007, 53, quoting Homer.

3. Ahura Mazda is a god of ancient Persia; Biema, of the Tiv in Nigeria; Chwezi, of the Banyoro in Uganda; Dakgipa, of the Garo in Bangladesh; Enuunap, of the Chuuk in Truk; Fundongthing, of the Lepcha in Sikkim; Great Spirit, of the Iroquois in the United States; Hokshi Togab, of the Assiniboine in Canada; Ijwala, of the Mataco in Argentina; Jehovah, of the ancient Hebrews; Kah-shu-goon-yah, of the Tlingit in the United States; Lata, of the Santa Cruz Indians in Polynesia; Mbori, of the Zande in the Central African Republic; Nkai, of the Maasai in Kenya; Osunduw, of the Rungus in Malaysia; Pab Dummat, of the Kuna in Panama; Quetzalcoatl, of the Toltecs in Mexico; Ra, of ancient Egypt; Sengalang Burong, of the Iban in Malaysia; Tirawa, of the Pawnee in the United States; Ugatame, of the Kapauku of Indonesia; Vodu, of the Ndyuka of French Guiana; Wiraqocha, of the Inca of Peru; Xi-He, of ancient China; Yurupari, of the Tupinamba of Brazil; and Zeus, of ancient Greece. A very useful resource for studying the gods is the online Human Relations Area Files at Yale University (www.yale.edu.hraf). Montaigne's *Essays*, book 2, chapter 12, is quoted in Robert J. Wenke and Deborah I. Olszewski, *Patterns in Prehistory* (New York: Oxford University Press, 2007), 315.
4. Annemarie De Waal Malefijt, *Religion and Culture* (New York: Macmillan, 1968), 153.
5. Nora Barlow, *The Autobiography of Charles Darwin, 1809–1882* (New York: Norton, 1958), 85; David Quammen, *The Reluctant Mr. Darwin* (New York: Norton, 2006), 42, 49; Paul H. Barrett, Peter J. Gautrey, Sandra Herbert et al., eds., *Charles Darwin's Notebooks, 1836–1844* (New York: Cambridge University Press, 1987), 291.
6. Charles Darwin, *The Descent of Man, and Selection in Relation to Sex* (London: John Murray, 1871), pt. 1, pp. 67, 68, and pt. 2, pp. 394–395, http://darwin- online.org.uk/content/frameset?viewtype=text&itemID=F937.2&pageseq=1.
7. Barlow, *The Autobiography of Charles Darwin*, 87, 90; Quammen, *The Reluctant Mr. Darwin*, 120. Darwin called himself an agnostic, but that was almost certainly to avoid offending his wife, who was a devout believer.
8. David J. Linden, *The Accidental Mind: How Brain Evolution Has Given Us Love, Memory, Dreams, and God* (Cambridge: Belknap Press of Harvard University Press, 2007), 28; Macdonald Critchley, *The Divine Banquet of the Brain, and Other Essays* (New York: Raven, 1979), 267. The claim that the human brain has ten times more glia than neurons has been questioned. That may be true for some brain areas but not others; see F. A. C. Azevedo, L. R. B. Carvalho, L. T. Grinberg et al., "Equal Numbers of Neuronal and Nonneuronal Cells Make the Human Brain an Isometrically Scaled-Up Primate Brain," *Journal of Comparative Neurology* 513 (2009): 532–541.
9. It should be noted that the Brodmann areas, which were originally defined on the basis of cellular differences seen under the microscope, do not necessarily correspond to functional areas. A few do, such as BA 3, the postcentral gyrus, which is the main sensory receptive area for the sense of touch. However, most Brodmann areas are involved

with multiple, different functions, especially those in areas known as being association cortices. In 2016 a new and more detailed system for numbering brain areas was introduced. It is based on images obtained using functional MRI, not on what the brain looks like under the microscope, as the Brodmann system does. Thus, the new numbering system will become standard for MRI, functional MRI, and other imaging studies, but it will not replace the Brodmann system, at least not in the near future.

10. M.-M. Mesulam, "Large-Scale Neurocognitive Networks and Distributed Processing for Attention, Language, and Memory," *Annals of Neurology* 28 (1990): 597–613; M.-M. Mesulam, "A Cortical Network for Directed Attention and Unilateral Neglect," *Annals of Neurology* 10 (1981): 309–325; M.-M. Mesulam, "From Sensation to Cognition," *Brain* 121 (1998): 1013–1052; J. K. Rilling, "Neuroscientific Approaches and Applications Within Anthropology," *Yearbook of Physical Anthropology* 51 (2008): 2–32. See also M. D. Fox, A. Z. Snyder, J. L. Vincent et al., "The Human Brain Is Intrinsically Organized Into Dynamic, Anticorrelated Functional Networks," *Proceedings of the National Academy of Sciences USA* 102 (2005): 9673–9678. The additional areas in the language network include the basal ganglia, inferior parietal lobule, middle temporal gyrus, inferior insula, and frontal cortex (Brodmann areas 6, 9, 45, and 47).
11. N. Gogtay, J. N. Giedd, L. Lusk et al., "Dynamic Mapping of Human Cortical Development During Childhood Through Early Adulthood," *Proceedings of the National Academy of Sciences USA* 101 (2004): 8174–8179.
12. Harry J. Jerison, *Evolution of the Brain and Intelligence* (New York: Academic, 1973), 9. Anthropologist Thomas Schoenemann has stated this principal as follows: "It is generally assumed . . . that more tissue translates into greater sophistication in neural processing in some way, which in turn suggests increased complexity of the behaviors mediated by that particular area (or areas)."
13. Gogtay et al., "Dynamic Mapping."
14. S. Wakana, H. Jiang, L. M. Nagae-Poetscher et al., "Fiber Track-Based Atlas of White Matter Human Anatomy," *Radiology* 230 (2004): 77–87; C. Lebel, L. Walker, A. Leemans et al., "Microstructural Maturation of the Human Brain from Childhood to Adulthood," *NeuroImage* 40 (2008): 1044–1055; W. Men, D. Falk, T. Sun et al., "The Corpus Callosum of Albert Einstein's Brain: Another Clue to His High Intelligence?" *Brain* 137 (2014): pt. 4, p. e268 (letter).
15. Stephen Jay Gould, *Ontogeny and Phylogeny* (Cambridge: Harvard University Press, 1977), 6; John C. Eccles, *Evolution of the Brain: Creation of the Self* (New York: Routledge, 1989), 203; D. Povinelli, "Reconstructing the Evolution of the Mind," *American Psychologist* 48 (1993): 493–509; S. T. Parker, "Comparative Developmental Evolutionary Biology, Anthropology, and Psychology," in *Biology, Brains, and Behavior*, ed. Sue Taylor Parker, Jonas Langer, and Michael L. McKinney (Santa Fe: School of American Research Press, 2000), 1–24, at 22.
16. J. W. Lichtman and Winfried Denk, "The Big and Small: Challenges of Imaging the Brain's Circuits," *Science* 334 (2011): 618–623. The Human Connectome Project, initiated by the National Institutes of Health in 2010, is currently mapping the brain's white matter connections and should markedly improve our understanding of these.

17. Charles Darwin, *Origin of Species* (New York: Collier, 1902), 126.
18. R. W. Scotland, "What Is Parallelism?," *Evolution and Development* 13 (2011): 214–227; David L. Smail, *On Deep History and the Brain* (Berkeley: University of California Press, 2008), 199. Parallel evolution differs from convergent evolution, which does not assume a common genetic ancestry, although the difference between the two has become obscured by the discovery of "deep homologies." For a discussion of this, see Gerhard Roth, *The Long Evolution of Brains and Minds* (New York: Springer, 2013), 37; and Stephen Jay Gould, *The Structure of Evolutionary Theory* (Cambridge: Harvard University Press, 2002), 1061–1089.
19. M. R. Leary and N. R. Buttermore, "The Evolution of the Human Self: Tracing the Natural History of Self-Awareness," *Journal for the Theory of Social Behaviour* 33 (2003): 365–404. See also Steven Mithen, *The Prehistory of the Mind: The Cognitive Origins of Art, Religion and Science* (London: Thames and Hudson, 1996) for a similar formulation.
20. M. Mesulam, "Brain, Mind, and the Evolution of Connectivity," *Brain and Cognition* 42 (2000): 4–6.

1. *HOMO HABILIS*

1. Stephen J. Gould, *Wonderful Life: The Burgess Shale and the Nature of History* (New York: Norton, 1989), 318. The forebrain gave rise to the diencephalon (thalamus and hypothalamus) and telencephalon (olfactory bulbs, hippocampus, amygdala, cingulate, basal ganglia, and cortex). The midbrain gave rise to the cerebral penducies and colliculi. The hindbrain gave rise to the medulla, pons, and cerebellum. A detailed analysis of the early brain development is found in Georg F. Striedter, *Principles of Brain Evolution* (Sunderland, MA: Sinauer, 2005).
2. Gould, *Wonderful Life*, 318, 44.
3. D. C. Van Essen and D. L. Dierker, "Surface-Based and Probabilistic Atlases of Primate Cerebral Cortex," *Neuron* 56 (2007): 209–225; Striedter, *Principles of Brain Evolution*, 287.
4. C. Zimmer, "A Twist on Our Ancestry," *New York Times*, October 29, 2013.
5. John S. Allen, *The Lives of the Brain: Human Evolution and the Organ of Mind* (Cambridge: Harvard University Press, 2009), 59–61; "Three of a Kind," *Economist*, September 10, 2005, 77.
6. Frederick L. Coolidge and Thomas Wynn, *The Rise of* Homo sapiens*: The Evolution of Modern Thinking* (New York: Wiley Blackwell, 2009), 87–90; R. L. Holloway, "The Casts of Fossil Hominid Brains," *Scientific American* 231 (1974): 106–115. Arguments about the endocasts of Australopithecus fossils have been going on for more than 30 years. A good summary of the data is in D. Falk, J. C. Redmond, J. Guyer et al., "Early Hominid Brain Evolution: A New Look at Old Endocasts," *Journal of Human Evolution* 38 (2000): 695–717; M. M. Skinner, N. B. Stephens, Z. J. Tsegai et al., "Human-Like Hand Use in Australopithecus Africanus," *Science* 347 (2015): 395–399;

M. Dominguez-Rodrigo, R. R. Pickering, and H. T. Bunn, "Configurational Approach to Identifying the Earliest Hominin Butchers," *Proceedings of the National Academy of Sciences USA* 107 (2010): 20929–20934; Coolidge and Wynn, *Rise of* Homo sapiens, 106.

7. Lewis Wolpert, *Six Impossible Things Before Breakfast: The Evolutionary Origins of Belief* (New York: Norton, 2006), 57; Michael C. Corballis, *From Hand to Mouth: The Origins of Language* (Princeton: Princeton University Press, 2002), 83–84; Jane Goodall, *The Chimpanzees of Gombe: Patterns of Behavior* (Cambridge: Harvard University Press, 1986), 535–545; Steven Mithen, *The Prehistory of the Mind: The Cognitive Origins of Art and Science* (London: Thames and Hudson, 1996), 96; Gerhard Roth, *The Long Evolution of Brains and Minds* (New York: Springer, 2013), 199.
8. Richard G. Klein and Blake Edgar, *The Dawn of Human Culture: A Bold New Theory on What Sparked the "Big Bang" of Human Consciousness* (New York: Wiley, 2002), 73–74; Mithen, *The Prehistory of the Mind*, 96–98.
9. Kenneth L. Feder, *The Past in Perspective: An Introduction to Human History* (Mountain View, CA: Mayfield, 2000), 81; D. Brown, "Arsenal Confirms Chimp's Ability to Plan, Study Says," *Washington Post*, March 10, 2009.
10. Nicholas Humphrey, *Consciousness Regained: Chapters in the Development of the Mind* (New York: Oxford University Press, 1984), 5, 48–49.
11. T. M. Preuss, "The Human Brain: Rewired and Running Hot," *Annals of the New York Academy of Sciences* 1225, supplement 1 (2011): E182–191; Richard Passingham, *What Is Special About the Human Brain?* (Oxford: Oxford University Press, 2008), 33; P. V. Tobias, "The Brain of Homo habilis: A New Level of Organization in Cerebral Evolution," *Journal of Human Evolution* 16 (1987): 741–761; Michael R. Rose, *Darwin's Spectre: Evolutionary Biology in the Modern World* (Princeton: Princeton University Press, 1998), 165.
12. Tobias, "The Brain of Homo habilis"; Holloway, "The Casts of Fossil Hominid Brains"; S. F. Witelson, D. L. Kigar, and T. Harvey, "The Exceptional Brain of Albert Einstein," *Lancet* 353 (1999): 2149–2153. In the brain collection of the Stanley Medical Research Institute, which is used for research on severe psychiatric disorders, there are 117 normal control brains (91 males and 26 females). The average weight is 1,472 grams, but they vary from 1,060 to 1,980 grams. It is assumed that 1 cubic centimeter of brain tissue weighs about 1 gram, so cubic centimeters and grams are roughly equivalent. R. E. Passingham, "The Origins of Human Intelligence," in *Human Origins*, ed. John R. Durant (Oxford: Clarendon, 1989), 123–136. Humans, however, do not have the largest brains for body size. The mouse lemur, a small primate, has a brain that is 3 percent of its body weight, whereas human brains are less than 2 percent of our body weight. Steve Jones, Robert Martin, David Pilbeam, eds., *The Cambridge Encyclopedia of Human Evolution* (Cambridge: Cambridge University Press, 1992), 107.
13. Tobias, "The Brain of Homo habilis."
14. R. E. Jung and R. J. Haier, "The Parieto-Frontal Integration Theory (P-FIT) of Intelligence: Converging Neuroimaging Evidence," *Behavioral and Brain Sciences* 30 (2007): 135–187. See also J. Gläscher, D. Tranel, L. K. Paul et al., "Lesion Mapping of

Cognitive Abilities Linked to Intelligence," *Neuron* 61 (2009): 681–691; J. Gläscher, D. Rudrauf, R. Colom et al., "Distributed Neural System for General Intelligence Revealed by Lesion Mapping," *Proceedings of the National Academy of Sciences USA* 107 (2010): 4705–4709; and A. K. Barbey, R. Colom, J. Solomon et al., "An Integrative Architecture for General Intelligence and Executive Function Revealed by Lesion Mapping," *Brain* 135 (2012): 1154–1164.

15. Jung and Haier, "The Parieto-Frontal Integration Theory."
16. Preuss, "The Human Brain"; M. L. McKinney, "Evolving Behavioral Complexity by Extending Development," in *Biology, Brains, and Behavior: The Evolution of Human Development*, ed. Sue Taylor Parker, Jonas Langer, and Michael L. McKinney (Santa Fe: School of American Research Press, 2000), 25–40, at 32; John C. Eccles, *Evolution of the Brain* (New York: Routledge, 1989), 42; Richard E. Passingham, *The Human Primate* (San Francisco: Freeman, 1982), 83; P. T. Schoenemann, "Evolution of the Size and Functional Areas of the Human Brain," *Annual Review of Anthropology* 35 (2006): 379–406.
17. K. Semendeferi, K. Teffer, D. P. Buxhoeveden et al., "Spatial Organization of Neurons in the Frontal Pole Sets Humans Apart from Great Apes," *Cerebral Cortex* 21 (2011): 1485–1497; S. Bludau, S. B. Eickhoff, H. Mohlberg et al., "Cytoarchitecture, Probability Maps and Functions of the Human Frontal Pole," *NeuroImage* 93 (2014): 260–275; K. Semendeferi, E. Armstrong, A. Schleicher et al., "Prefrontal Cortex in Humans and Apes: A Comparative Study of Area 10," *American Journal of Physical Anthropology* 114 (2001): 224–241; R. Muhammad, J. D. Wallis, and E. K. Miller, "A Comparison of Abstract Rules in the Prefrontal Cortex, Premotor Cortex, Inferior Temporal Cortex, and Striatum," *Journal of Cognitive Neuroscience* 18 (2006): 974–989; P. J. Brasted and S. P. Wise, "Comparison of Learning-Related Neuronal Activity in the Dorsal Premotor Cortex and Striatum," *European Journal of Neuroscience* 19 (2004): 721–740; J. M. Fuster, "Frontal Lobe and Cognitive Development," *Journal of Neurocytology* 31 (2002): 373–385; J. Jonides, E. E. Smith, R. A. Koeppe et al., "Spatial Working Memory in Humans as Revealed by PET (Letter)," *Nature* 363 (1993): 623–625.
18. A. E. Cavanna and M. R. Trimble, "The Precuneus: A Review of Its Functional Anatomy and Behavioural Correlates," *Brain* 129 (2006): 564–583; Witelson et al., "The Exceptional Brain of Albert Einstein"; W. Men, D. Falk, T. Sun et al., "The Corpus Callosum of Albert Einstein's Brain: Another Clue to His High Intelligence?," *Brain* 137 (2014): pt. 4, p. e268 (letter).
19. N. Makris, D. N. Kennedy, S. McInerney et al., "Segmentation of Subcomponents Within the Superior Longitudinal Fascicle in Humans: A Quantitative, In Vivo, DT-MRI Study," *Cerebral Cortex* 15 (2005): 854–869; T. Sakai, A. Mikami, M. Tomonaga et al., "Differential Prefrontal White Matter Development in Chimpanzees and Humans," *Current Biology* 21 (2011): 1397–1402; J. S. Schneiderman, M. S. Buchsbaum, M. M. Haznedar et al., "Diffusion Tensor Anistrophy in Adolescents and Adults," *Neuropsychobiology* 55 (2007): 96–111; J. Zhang, A. Evans, L. Hermoye et al., "Evidence of Slow Maturation of the Superior Longitudinal Fasciculus in Early Childhood by

Diffusion Tensor Imaging," *NeuroImage* 38 (2007): 239–247; G. Roth and U. Dicke, "Evolution of the Brain and Intelligence," *Trends in Cognitive Science* 29 (2005): 250–257.

20. Ian Tattersall, *Becoming Human: Evolution and Human Uniqueness* (New York: Harcourt Brace, 1998), 194. Striedter in *Principles of Brain Evolution* states it as follows: "As [brain] regions become disproportionately large, they tend to 'invade' regions that they did not innervate ancestrally" (11); P. V. Tobias, "Recent Advances in the Evolution of the Hominids with Special Reference to Brain and Speech," in *Recent Advances in the Evolution of Primates*, ed. Carlos Chagas (Vatican City: Pontificiae Academiae Scientiarum Scripta Varia 50, 1983), 85–140. Tobias was quoting N. Geschwind, "Disconnexion Syndromes in Animals and Nan," *Brain* 88 (1965): 237–294, and N. W. Ingalls, "The Parietal Region in the Primate Brain," *Journal of Comparative Neurology* 24 (1914): 291–341. See also MacDonald Critchley, *The Parietal Lobes* (New York: Hafner, 1969), 16; M.-M. Mesulam, "A Cortical Network for Directed Attention and Unilateral Neglect," *Annals of Neurology* 10 (1981): 309–325; and Striedter, *Principles of Brain Evolution*, 327. The inferior parietal lobule consists of areas 39 and 40 in the Brodmann system of classification.
21. R. I. M. Dunbar, "The Social Brain Hypothesis and Its Implications for Social Evolution," *Annals of Human Biology* 36 (2009): 562–572; M. Balter, "Why Are Our Brains So Big?," *Science* 338 (2012): 33–34.

2. *HOMO ERECTUS*

1. F. Spoor, M. G. Leakey, P. N. Gathogo et al., "Implications of New Early *Homo* Fossils from Ileret, East of Lake Turkana, Kenya (Letter)," *Nature* 448 (2007): 688–691; J. N. Wilford, "New Fossils Indicate Early Branching of Human Family Tree," *New York Times*, August 9, 2012. We know remarkably little about which species evolved from which, since we still have so few fossils. Archeologists argue endlessly about such issues, but it is like arguing about what a five-hundred-piece jigsaw puzzle will look like when only twenty-seven of the pieces are yet available.
2. Andrew Shyrock and Daniel Lord Smail, *Deep History: The Architecture of Past and Present* (Berkeley: University of California Press, 2011), 69–70.
3. Kenneth L. Feder, *The Past in Perspective: An Introduction to Human History* (Mountain View, CA: Mayfield, 2000), 120–121. The stone tools of *Homo erectus* are usually classified as Acheulean, in contrast to the Oldowan (named after Olduvai Gorge) tools of *Homo habilis*. It should be pointed out that, although we call the tools "handaxes," we really do not know how they were used. See R. G. Klein, "Archeology and the Evolution of Human Behavior," *Evolutionary Anthropology* 9 (2000): 17–36.
4. Frederick L. Coolidge and Thomas Wynn, *The Rise of* Homo sapiens*: The Evolution of Modern Thinking* (New York: Wiley Blackwell, 2009), 151; Z. Zorich, "The First Spears," *Archaeology*, March–April 2013, 16; M. Balter, "The Killing Ground," *Science* 344 (2014): 1080–1083.

5. A. Gibbons, "Food for Thought: Did the First Cooked Meals Help Fuel the Dramatic Evolutionary Expansion of the Human Brain," *Science* 316 (2007): 1558–1560; J. Gorman, "Chimps Would Cook If Given Chance, Research Says," *New York Times*, June 3, 2015. See also Richard Wrangham, *Catching Fire: How Cooking Made Us Human* (New York: Basic, 2009).
6. See, for example, Terrence C. Deacon, *The Symbolic Species: The Co-Evolution of Language and the Brain* (New York: Norton, 1997).
7. Merlin Donald, *Origins of the Modern Mind* (Cambridge: Harvard University Press, 1991), 112.
8. M. Lewis, "Myself and Me," in *Self-Awareness in Animals and Humans: Developmental Perspectives*, ed. Sue Taylor Parker, Robert W. Mitchell, and Maria L. Boccia (New York: Cambridge University Press, 1994), 20–34.
9. B. Amsterdam, "Mirror Self-Image Reactions Before Age Two," *Developmental Psychobiology* 5 (1972): 297–305.
10. J. R. Anderson, "To See Ourselves as Others See Us: A Response to Mitchell," *New Ideas in Psychology* 11 (1993): 339–346; J. R. Anderson, "The Development of Self-Recognition: A Review," *Developmental Psychobiology* 17 (1984): 37–49; M. Lewis and J. Brooks-Gunn, "Toward a Theory of Social Cognition: The Development of Self," in *Social Interaction and Communication During Infancy*, ed. Ina C. Užgiris (Washington, DC: Jossey-Bass, 1979), 1–20; L. Mans, D. Cicchetti, and L. A. Sroufe, "Mirror Reactions of Down's Syndrome Infants and Toddlers: Cognitive Underpinnings of Self-Recognition," *Child Development* 49 (1978): 1247–1250; G. Dawson and F. C. McKissick, "Self-Recognition in Autistic Children," *Journal of Autism and Developmental Disorders* 14 (1984): 383–394; C. J. Neuman and S. D. Hill, "Self-Recognition and Stimulus Preference in Autistic Children," *Developmental Psychobiology* 11 (1978): 571–578.
11. A. D. Craig, "How Do You Feel—Now? The Anterior Insula and Human Awareness," *Nature Reviews Neuroscience* 10 (2009): 59–70; Antonio Damasio, "The Person Within," *Nature* 423 (2003): 227; A. D. Craig, "The Sentient Self," *Brain Structure and Function* 214 (2010): 563–577; G. Gallup, "Self-Awareness and the Emergence of Mind in Primates," *American Journal of Primatology* 2 (1982): 237–248.
12. S. D. Hill and C. Tomlin, "Self-Recognition in Retarded Children," *Child Development* 52 (1981): 145–150; T. F. Pechacek, K. F. Bell, C. C. Cleland et al., "Self-Recognition in Profoundly Retarded Males," *Bulletin of the Psychonomic Society* 1 (1973): 328–330; L. P. Harris, "Self-Recognition Among Institutionalized Profoundly Retarded Males: A Replication," *Bulletin of the Psychonomic Society* 9 (1977): 43–44.
13. E. F. Torrey, "Schizophrenia and the Inferior Parietal Lobule," *Schizophrenia Research* 97 (2007): 215–225; D. Simeon, O. Guralnik, E. A. Hazlett et al., "Feeling Unreal: A PET Study of Depersonalization Disorder," *American Journal of* Psychiatry 157 (2000): 1782–1788; F. Biringer, J. R. Anderson, and D. Strubel, "Self-Recognition in Senile Dementia," *Experimental Aging Research* 14 (1988): 177–180; F. Biringer and J. R. Anderson, "Self-Recognition in Alzheimer's Disease: A Mirror and Video Study," *Journal of Gerontology* 47 (1992): P385–P388; E. H. Rubin, W. C. Drevets, and W. J. Burke, "The Nature of Psychotic Symptoms in Senile Dementia of the Alzheimer

Type," *Journal of Geriatric Psychiatry and Neurology* 1 (1988): 16–20; Todd E. Feinberg, *Altered Egos: How the Brain Creates the Self* (New York: Oxford University Press, 2001), 73; L. K. Gluckman, "A Case of Capgras Syndrome," *Australian and New Zealand Journal of Psychiatry* 2 (1968): 39–43.

14. G. G. Gallup Jr., "Chimpanzees: Self-Recognition," *Science* 167 (1970): 86–87.
15. Gerhard Roth, *The Long Evolution of Brains and Minds* (New York: Springer, 2013), 210; H. L. W. Miles, "Me Chantek: The Development of Self-Awareness in a Signing Orangutan," in Parker, Mitchell, and Boccia, *Self-Awareness in Animals and Humans*, 254–272; Michael Lewis and Jeanne Brooks-Gunn, *Social Cognition and the Acquisition of Self* (New York: Plenum, 1979), 182.
16. H. Prior, A. Schwarz, and O. Güntürkün, "Mirror-Induced Behavior in the Magpie (*Pica pica*): Evidence of Self-Recognition," *PLoS Biology* 6 (2008): e202; J. M. Plotnik, F. B. M. de Waal, and D. Reiss, "Self-Recognition in an Asian Elephant," *Proceedings of the National Academy of Sciences USA* 103 (2006): 17063–17057; D. Reiss and L. Marino, "Mirror Self-Recognition in the Bottlenose Dolphin: A Case of Cognitive Convergence," *Proceedings of the National Academy of Sciences USA* 98 (2001): 5937–5942; Nicholas Humphrey, *The Inner Eye: Social Intelligence in Evolution* (New York: Oxford University Press, 2002), 84.
17. Roth, *The Long Evolution of Brains and Minds*, 210; C. W. Hyatt and W. D. Hopkins, "Self-Awareness in Bonobos and Chimpanzees: A Comparative Perspective," in Parker, Mitchell, and Boccia, *Self-Awareness in Animals and Humans*, 248–253; Miles, "Me Chantek"; F. B. M. de Waal, M. Dindo, A. Freeman et al., "The Monkey in the Mirror: Hardly a Stranger," *Proceedings of the National Academy of Sciences USA* 102 (2005): 11140–11147. Self-awareness may have evolved independently several times. Such things happen in evolution; for example, it has been said that the eye evolved independently at least 40 times in human evolution. See Steven Pinker, *The Language Instinct* (New York: HarperCollins, 1995), 349; Richard Dawkins, *The Ancestor's Tale: A Pilgrimage to the Dawn of Evolution* (Boston: Houghton Mifflin, 2004), 589; E. Pennisi, "Mining the Molecules That Made Our Mind, *Science* 313 (2006): 1908–1911; H. E. Hoekstra and T. Price, "Parallel Evolution Is in the Genes," *Science* 303 (2004): 1779–1781; M. R. Leary and N. R. Buttermore, "The Evolution of the Human Self: Tracing the Natural History of Self-Awareness," *Journal for the Theory of Social Behaviour* 33 (2003): 365–404 (see also Steven Mithen, *The Prehistory of the Mind: The Cognitive Origins of Art, Religion and Science* [London: Thames and Hudson, 1996] for a similar formulation); Richard G. Klein and Blake Edgar, *The Dawn of Human Culture: A Bold New Theory on What Sparked the "Big Bang" of Human Consciousness* (New York: Wiley, 2002), 8; John Hawks, Eric T. Wang, Gregory M. Cochran et al., "Recent Acceleration of Human Adaptive Evolution," *Proceedings of the National Academy of Sciences USA* 104 (2007): 20753–20758; see also Patrick Evans, Sandra L. Gilbert, Nitzan Mekel-Bobrov et al., "*Microcephalin*, a Gene Regulating Brain Size, Continues to Evolve Adaptively in Humans," *Science* 309 (2005): 1717–1720.
18. S. T. Parker, "A Social Selection Model for the Evolution and Adaptive Significance of Self-Conscious Emotions," in *Self-Awareness: Its Nature and Development*,

ed. Michael Ferrari and Robert J. Sternberg (New York: Guilford, 1998), 108–136; Ian Tattersall, *Becoming Human: Evolution and Human Uniqueness* (New York: Harcourt Brace, 1998), 48; Raymond Tallis, *The Kingdom of Infinite Space: A Fantastical Journey Around Your Head* (New Haven: Yale University Press, 2008), 220–221.

19. Feder, *The Past in Perspective*, 106; Donald, *Origins of the Modern Mind*, 113. For a similar analysis of one of the most recently found *Homo erectus* skulls, see X. Wu, L. A. Schepartz, and W. Liu, "A New *Homo erectus* (Zhoukoudian V) Brain Endocast from China," *Proceedings of the Royal Society B* 277 (2009): 337–344; Coolidge and Wynn, *The Rise of* Homo Sapiens, 114.
20. John S. Allen, *The Lives of the Brain: Human Evolution and the Organ of Mind* (Cambridge: Harvard University Press, 2009), 98; Craig, "How Do You Feel—Now?"
21. C. Lebel, L. Walker, A. Leemans et al., "Microstructural Maturation of the Human Brain from Childhood to Adulthood," *NeuroImage* 40 (2008).
22. D. T. Stuss, "Disturbance of Self-Awareness After Frontal System Damage," in *Awareness of Deficit After Brain Injury: Clinical and Theoretical Issues*, ed. George P. Prigatano and Daniel L. Schacter (New York: Oxford University Press, 1991), 63–83, 65, 68; K. P. Wylie and J. R. Tregallas, "The Role of the Insula in Schizophrenia," *Schizophrenia Research* 123 (2010): 93–104; Craig, "How Do You Feel—Now?"
23. K. Zilles, "Architecture of the Human Cerebral Cortex," in *The Human Nervous System*, ed. George Paxinos and Juergen K. Mai, 2nd ed. (Amsterdam: Elsevier, 2004), 997–1042. Inferior parietal lobule function is, in fact, extraordinarily complex and diverse. For older reviews, see MacDonald Critchley, *The Parietal Lobes* (New York: Hafner, 1969); and D. Denny-Brown and R. A. Chambers, "The Parietal Lobe and Behavior," in *The Brain and Human Behavior*, ed. Harry C. Solomon, Stanley Cobb, and Wilder Penfield (Baltimore: Williams and Wilkins, 1958), 35–117.
24. T. W. Kjaer, M. Nowak, and H. C. Lou, "Reflective Self-Awareness and Conscious States: PET Evidence for a Common Midline Parietofrontal Core," *NeuroImage* 17 (2002): 1080–1086; P. Ruby and J. Decety, "Effect of Subjective Perspective Taking During Simulation of Action: A PET Investigation of Agency," *Nature Neuroscience* 4 (2001): 546–550; L. Q. Uddin, J. T. Kaplan, I. Molnar-Szakacs et al., "Self-Face Recognition Activates a Frontoparietal 'Mirror' Network in the Right Hemisphere: An Event-Related fMRI Study," *NeuroImage* 25 (2005): 926–935; H. C. Lou, B. Luber, M. Crupain et al., "Parietal Cortex and Representation of the Mental Self," *Proceedings of the National Academy of Sciences USA* 101 (2004): 6827–6832; S. M. Platek, J. W. Loughead, R. C. Gur et al., "Neural Substrates for Functionally Discriminating Self-Face from Personally Familiar Faces," *Human Brain Mapping* 27 (2006): 91–98; Simeon et al., "Feeling Unreal."
25. C. Butti, M. Santos, N. Uppal et al., "Von Economo Neurons: Clinical and Evolutionary Perspectives," *Cortex* 49 (2013): 312–326; J. M. Allman, N. A. Tetreault, A. Y. Hakeem et al., "The Von Economo Neurons in Frontoinsular and Anterior Cingulate Cortex in Great Apes and Humans," *Brain Structure and Function* 214 (2010): 495–517.
26. F. Cauda, G. C. Geminiani, and A. Vercelli, "Evolutionary Appearance of Von Economo's Neurons in the Mammalian Cerebral Cortex," *Frontiers in Human*

Neuroscience 8 (2014): 104; C. Fajardo, M. I. Escobar, E. Buriticá et al., "Von Economo Neurons Are Present in the Dorsolateral (Dysgranular) Prefrontal Cortex of Humans," *Neuroscience Letters* 435 (2008): 215–218; C. Butti, C. C. Sherwood, A. Y. Hakeem et al., "Total Number and Volume of Von Economo Neurons in the Cerebral Cortex of Cetaceans," *Journal of Comparative Neurology* 515 (2009): 243–259; Allman et al., "The Von Economo Neurons."

27. V. E. Sturm, H. J. Rosen, S. Allison et al., "Self-Conscious Emotion Deficits in Frontotemporal Lobar Degeneration," *Brain* 129 (2006): 2508–2516; W. W. Seeley, D. A. Carlin, J. M. Allman et al., "Early Frontotemporal Dementia Targets Neurons Unique to Apes and Humans, *Annals of Neurology* 60 (2006): 660–667.
28. Allman et al., "The Von Economo Neurons"; J. Allman, Atiya Hakeem, and K. Watson, "Two Phylogenetic Specializations in the Human Brain," *Neuroscientist* 8 (2002): 335–346; J. M. Allman, N. A. Tetreault, A. Y. Hakeem et al., "The Von Economo Neurons in the Frontoinsular and Anterior Cingulate Cortex," *Annals of the New York Academy of Sciences* 1225 (2011): 59–71.

3. ARCHAIC *HOMO SAPIENS* (NEANDERTALS)

1. N. Wade, "Genetic Data and Fossil Evidence Tell Differing Tales of Human Origins," *New York Times*, July 27, 2012; J.-J. Hublin, "How to Build a Neandertal," *Science* 344 (2014): 1338–1339; A. Gibbons, "Who Were the Denisovans?," *Science* 333 (2011): 1084–1087; E. Culotta, "Likely Hobbit Ancestors Lived 600,000 Years Earlier," *Science* 352 (2016): 1260–1261; A. Gibbons, "A Crystal-Clear View of an Extinct Girl's Genome," *Science* 337 (2012): 1028–1029; M. Meyer, M. Kircher, M.-T. Gansauge et al., "A High-Coverage Genome Sequence from an Archaic Denisovan Individual," *Science* 338 (2012): 222–226; A. Cooper and C. B. Stringer, "Did Denisovans Cross Wallace's Line?," *Science* 342 (2013): 321–323.
2. A. W. Briggs, J. M. Good, R. E. Green et al., "Targeted Retrieval and Analysis of Five Neandertal mtDNA Genomes," *Science* 325 (2009): 318–320.
3. Richard G. Klein and Blake Edgar, *The Dawn of Human Culture: A Bold New Theory on What Sparked the "Big Bang" of Human Consciousness* (New York: Wiley, 2002), 272.
4. Brian Fagan, *Cro-Magnon: How the Ice Age Gave Birth to the First Modern Humans* (New York: Bloomsbury, 2010), 47.
5. K. Bouton, "If Cave Men Told Jokes, Would Humans Laugh?," *New York Times*, December 28, 2011; D. S. Adler, K. N. Wilkinson, S. Blockley et al., "Early Levallois Technology and the Lower to Middle Paleolithic Transition in the Southern Caucasus," *Science* 345 (2014): 1609–1613; Carl Zimmer, *Evolution: The Triumph of an Idea* (New York: HarperCollins, 2001), 301; M. Soressi, S. P. McPherron, M. Lenoir et al., "Neandertals Made the First Specialized Bone Tools in Europe," *Proceedings of the National Academy of Sciences USA* 110 (2013): 14186–14190. See also Christopher Stringer and Clive Gamble, *In Search of the Neanderthals* (London: Thames and

Hudson, 1993). The method used by Neandertals for making stone tools is usually referred to as the Levallois technique.

6. Fagan, *Cro-Magnon*, 80; D. Bickerton, "From Protolanguage to Language," in *The Speciation of Modern* Homo sapiens, ed. Tim J. Crow (Oxford: Oxford University Press, 2002), 103–120.
7. W. Roebroeks, M. J. Sier, T. K. Nielsen et al., "Use of Red Ochre by Early Neandertals," *Proceedings of the National Academy of Sciences USA* 109 (2012): 1889–1894; J. Zilhão, D. E. Angelucci, E. Badal-Garcia et al., "Symbolic Use of Marine Shells and Mineral Pigments by Iberian Neandertals," *Proceedings of the National Academy of Sciences USA* 107 (2010): 1023–1028; M. Peresani, M. Vanhaeren, E. Quaggiotto et al., "An Ochered Fossil Marine Shell from the Mousterian of Fumane Cave, Italy," *PLoS ONE* 8 (2013): e68572; E. Morin and V. Laroulandie, "Presumed Symbolic Use of Diurnal Raptors by Neanderthals," *PLoS ONE* 7 (2012): e32856; C. Finlayson, K. Brown, R. Blasco et al., "Birds of a Feather: Neanderthal Exploitation of Raptors and Corvids," *PLoS ONE* 7 (2012): e45927; M. Peresani, I. Fiore, M. Gala et al., "Late Neandertals and the Intentional Removal of Feathers as Evidenced from Bird Bone Taphonomy at Fumane Cave 44 ky BP., Italy," *Proceedings of the National Academy of Sciences USA* 108 (2011): 3888–3893; J. Rodriguez-Vidal, F. d'Errico, F. G. Pacheco et al., "A Rock Engraving Made by Neanderthals in Gibraltar," *Proceedings of the National Academy of Sciences USA* 111 (2014): 13301–13306; M. Romandini, M. Peresani, V. Laroulandie et al., "Convergent Evidence of Eagle Talons Used by Late Neanderthals in Europe: A Further Assessment on Symbolism," *PLoS ONE* 9 (2014): e101278.
8. Stringer and Gamble, *In Search of the Neanderthals*, 94; Kenneth L. Feder, *The Past in Perspective: An Introduction to Human History* (Mountain View, CA: Mayfield, 2000), 161; Chris Stringer, *Lone Survivors: How We Came to Be the Only Humans on Earth* (New York: Times, 2012), 153–154; Robert J. Wenke and Deborah I. Olszewski, *Patterns in Prehistory: Mankind's First Three Million Years* (Oxford: Oxford University Press, 2007), 162; Gregory Curtis, *The Cave Painters: Probing the Mysteries of the World's First Artists* (New York: Anchor, 2006), 34.
9. A. Belfer-Cohen and E. Hovers, "In the Eye of the Beholder: Mousterian and Natufian burials in the Levant," *Current Anthropology* 33 (1992): 463–471; Ian Tattersall, *Becoming Human: Evolution and Human Uniqueness* (New York: Harcourt Brace, 1998), 161, 162–163; Fagan, *Cro-Magnon*, 77.
10. R. N. Spreng, R. A. Mar, and S. N. Kim, "The Common Neural Basis of Autobiographical Memory, Prospection, Navigation, Theory of Mind, and the Default Mode: A Quantitative Meta-Analysis," *Journal of Cognitive Neuroscience* 21 (2009): 489–510; Nicholas Humphrey, *The Inner Eye: Social Intelligence in Evolution* (New York: Oxford University Press, 2002), 71.
11. C. D. Frith, "Schizophrenia and Theory of Mind (Editorial)," *Psychological Medicine* 34 (2004): 385–389.
12. D. J. Povinelli and C. G. Prince, "When Self Met Other," in *Self-Awareness: Its Nature and Development*, ed. Michael Ferrari and Robert J. Sternberg (New York: Guilford, 1998), 62; C. D. Frith and U. Frith, "Interacting Minds—a Biological Basis,"

Science 286 (1999): 1692–1695; J. I. M. Carpendale and C. Lewis, "Constructing an Understanding of Mind: The Development of Children's Social Understanding Within Social Interaction," *Behavioral and Brain Sciences* 27 (2004): 79–151. See also Robin Dunbar, *The Human Story: A New History of Mankind's Evolution* (London: Faber and Faber, 2004), 43. Additional evidence that the acquisition of a theory of mind can be improved by training comes from studies showing that reading literary fiction improves theory of mind skills in adults; see D. C. Kidd and E. Castano, "Reading Literary Fiction Improves Theory of Mind," *Science* 342 (2013): 377–380.

13. A. Y. Hakeem, C. C. Sherwood, C. J. Bonar et al., "Von Economo Neurons in the Elephant Brain," *Anatomical Record* 292 (2009): 242–248.
14. A. Jolly, "The Social Origin of Mind (Book Review)," *Science* 317 (2007): 1326–1327.
15. See, for example, Jane Goodall, *The Chimpanzees of Gombe: Patterns of Behavior* (Cambridge: Harvard University Press, 1986), 36–38, 578–583; and Barbara J. King, *Evolving God: A Provocative View on the Origins of Religion* (New York: Doubleday, 2007), 36.
16. Zimmer, *Evolution*, 271; Dunbar, *The Human Story*, 59; M. Tomasello, J. Call, and B. Hare, "Chimpanzees Understand Psychological States—the Question Is Which Ones and to What Extent," *Trends in Cognitive Sciences* 7 (2003): 153–156; Povinelli and Prince, "When Self Met Other," 93. For useful summaries of this debate, see also D. J. Povinelli and J. M. Bering, "The Mentality of Apes Revisited," *Current Directions in Psychological Science* 11 (2002): 115–119; D. J. Povinelli and T. M. Preuss, "Theory of Mind: Evolutionary History of a Cognitive Specialization," *Trends in Neurosciences* 18 (1995): 418–424; D. C. Penn and D. J. Povinelli, "On the Lack of Evidence That Non-Human Animals Possess Anything Remotely Resembling a 'Theory of Mind,'" *Philosophical Transactions of the Royal Society* 362 (2007): 731–744; J. B. Silk, S. F. Brosnan, J. Vonk et al., "Chimpanzees Are Indifferent to the Welfare of Unrelated Group Members," *Nature* 437 (2005): 1357–1359. See also a recent discussion of this issue in Thomas Suddendorf, *The Gap: The Science of What Separates Us from Other Animals* (New York: Basic, 2013), 126–132.
17. Richard M. Restak, *The Modular Brain* (New York: Touchstone, 1994), 107.
18. A. M. Leslie, "The Theory of Mind Impairment in Autism: Evidence for a Modular Mechanism of Development?," in *Natural Theories of Mind: Evolution, Development and Simulation of Everyday Mindreading*, ed. Andrew Whiten (Oxford: Basil Blackwell, 1991), 63–77; Simon Baron-Cohen, *Mindblindness: An Essay on Autism and Theory of Mind* (Cambridge: MIT Press, 1997).
19. Y. Yang, A. L. Glenn, and A. Raine, "Brain Abnormalities in Antisocial Individuals: Implications for the Law," *Behavioral Sciences and the Law* 26 (2008): 65–83; M. Macmillan, "Inhibition and the Control of Behavior: From Gall to Freud via Phineas Gage and the Frontal Lobes," *Brain and Cognition* 19 (1992): 72–104; E. L. Hutton, "Personality Changes After Leucotomy," *Journal of Mental Science* 93 (1947): 31–42; Jack El-Hai, *The Lobotomist* (New York: Wiley, 2005), 168.
20. C. B. Stringer, "Evolution of Early Humans," in *The Cambridge Encyclopedia of Human Evolution*, ed. Steve Jones, Robert D. Martin, and David R. Pilbeam (Cambridge:

Cambridge University Press, 1992), 245; MacDonald Critchley, *The Parietal Lobes* (New York: Haffner, 1969), 54.

21. Percival Bailey and Gerhardt von Bonin, *The Isocortex of Man* (Champagne: University of Illinois Press, 1951), 218; R. M. Carter, D. L. Bowling, C. Reeck et al., "A Distinct Role of the Temporal-Parietal Junction in Predicting Socially Guided Decisions," *Science* 337 (2012): 109–111; G. D. Pearlson, "Superior Temporal Gyrus and Planum Temporale in Schizophrenia: A Selective Review," *Progress in Neuro-Psychopharmacology and Biological Psychiatry* 21 (1997): 1203–1229.
22. N. Makris, D. N. Kennedy, S. McInerney et al., "Segmentation of Subcomponents Within the Superior Longitudinal Fascicle in Humans: A Quantitative, In Vivo, DT-MRI Study," *Cerebral Cortex* 15 (2005); J. K. Rilling, M. F. Glasser, T. M. Preuss et al., "The Evolution of the Arcuate Fasciculus Revealed with Comparative DTI," *Nature Neuroscience* 11 (2008): 426–428.
23. R. Saxe and A. Wexler, "Making Sense of Another Mind: The Role of the Right Temporo-Parietal Junction," *Neuropsychologia* 43 (2005): 1391–1399; J. S. Rabin, A. Gilboa, D. T. Stuss et al., "Common and Unique Neural Correlates of Autobiographical Memory and Theory of Mind," *Journal of Cognitive Neuroscience* 22 (2010): 1095–1111; J. Decety and J. Grèzes, "The Power of Stimulation: Imagining One's Own and Other's Behavior," *Brain Research* 1079 (2006): 4–14; Martin Brüne and Ute Brüne-Cohrs, "Theory of Mind—Evolution, Ontogeny, Brain Mechanisms and Psychopathology," *Neuroscience and Biobehavioral Reviews* 30 (2006): 437–455. See also R. Saxe and N. Kanwisher, "People Thinking About People: The Role of the Temporo-Parietal Junction in 'Theory of Mind,'" *NeuroImage* 19 (2003): 1835–1842.
24. John S. Allen, *The Lives of the Brain: Human Evolution and the Organ of Mind* (Cambridge: Harvard University Press, 2009), 97; Spreng et al., "The Common Neural Basis"; L. Carr, M. Iacoboni, M.-C. Dubeau et al., "Neural Mechanisms of Empathy in Humans: A Relay from Neural Systems for Imitation to Limbic Areas," *Proceedings of the National Academy of Sciences USA* 100 (2003): 5497–5502; K. N. Ochsner, J. Zaki, J. Hanelin et al., "Your Pain or Mine? Common and Distinct Neural Systems Supporting the Perception of Pain in Self and Other," *Social Cognitive and Affective Neuroscience* 3 (2008): 144–160.
25. D. Falk, C. Hildebolt, K. Smith et al., "The Brain of LB1, *Homo florensiensis*," *Science* 308 (2005): 242–245; C. D. Frith and U. Frith, "Interacting Minds—a Biological Basis," *Science* 286 (1999): 1692–1695; C. D. Frith and U. Frith, "The Neural Basis of Mentalizing," *Neuron* 50 (2006): 531–534; P. C. Fletcher, F. Happé, U. Frith et al., "Other Minds in the Brain: A Functional Imaging Study of 'Theory of Mind' in Story Comprehension," *Cognition* 57 (1995): 109–128; D. T. Stuss, G. G. Gallup Jr., and M. P. Alexander, "The Frontal Lobes Are Necessary for 'Theory of Mind,'" *Brain* 124 (2001): 279–286.
26. G. Rizzolatti and L. Craighero, "The Mirror-Neuron System," *Annual Review of Neuroscience* 27 (2004): 169–192; Decety, and Grèzes, "The Power of Stimulation"; Andrew Shryock and Daniel L. Smail, *Deep History* (Berkeley: University of California Press, 2011), 63.
27. Jesse Bering, *The Belief Instinct: The Psychology of Souls, Destiny, and the Meaning of Life* (New York: Norton, 2011), 190; Ara Norenzayan, *Big Gods: How Religion*

Transformed Cooperation and Conflict (Princeton: Princeton University Press, 2013); Dominic Johnson, *God Is Watching You: How the Fear of God Makes Us Human* (New York: Oxford University Press, 2016).

28. Bering, *The Belief Instinct*, 190, 192.
29. W. M. Gervais, "Perceiving Minds and Gods: How Mind Perception Enables, Constrains, and Is Triggered by Belief in Gods," *Perspectives on Psychological Science* 8 (2013): 380–394.
30. D. Kapogiannis, A. K. Barbey, M. Su et al., "Cognitive and Neural Foundations of Religious Belief," *Proceedings of the National Academy of Sciences USA* 106 (2009): 4876–4881.

4. EARLY *HOMO SAPIENS*

1. J. R. Stewart and C. B. Stringer, "Human Evolution out of Africa: The Role of Refugia and Climate Change," *Science* 335 (2012): 1317–1321; Chris Stringer, *Lone Survivors: How We Came to Be the Only Humans on Earth* (New York: Times, 2012), 130.
2. V. Mourre, P. Villa, and C. S. Henshilwood, "Early Use of Pressure Flaking on Lithic Artifacts at Blombos Cave, South Africa," *Science* 330 (2010): 659–662; P. Mellars, "Archeology and the Origins of Modern Humans: European and African Perspectives," in *The Speciation of Modern* Homo sapiens, ed. Tim J. Crow (Oxford: Oxford University Press, 2002), 37, 39; C. S. Henshilwood, J. C. Sealy, R. Yates et al., "Blombos Cave, Southern Cape, South Africa: Preliminary Report on the 1992–1999 Excavations of the Middle Stone Age Levels," *Journal of Archaeological Science* 28 (2001): 421–448; L. Wadley, C. Sievers, M. Bamford et al., "Middle Stone Age Bedding Construction and Settlement Patterns at Sibudu, South Africa," *Science* 334 (2011): 1388–1391; M. Balter, "South African Cave Slowly Shares Secrets of Human Culture," *Science* 332 (2011): 1260–1261; S. McBrearty and A. S. Brooks, "The Revolution That Wasn't: A New Interpretation of the Origin of Modern Human Behavior," *Journal of Human Evolution* 39 (2000): 453–563; M. Lombard, "Quartz-Tipped Arrows Older Than 60 Ka: Further Use-Trace Evidence from Sibudu, KwaZulu-Natal, South Africa," *Journal of Archaeological Science* 38 (2011): 1918–1930.
3. Wadley et al., "Middle Stone Age Bedding Construction."
4. M. Balter, "First Jewelry? Old Shell Beads Suggest Early Use of Symbols," *Science* 312 (2006): 173; M. Vanhaeren, F. d'Errico, C. Stringer et al., "Middle Paleolithic Shell Beads in Israel and Algeria," *Science* 312 (2006): 1785–1788; C. S. Henshilwood, F. d'Errico, K. L. van Niekerk et al., "A 100,000-Year-Old Ochre-Processing Workshop at Blombos Cave, South Africa," *Science* 334 (2011): 219–222; F. d'Errico, M. Vanhaeren, N. Barton et al., "Additional Evidence on the Use of Personal Ornaments in the Middle Paleolithic of North America," *Proceedings of the National Academy of Sciences USA* 106 (2009): 16051–16056; E. A. Powell, "In Style in the Stone Age," *Archaeology* (July–August 2013): 18.
5. C. S. Henshilwood, F. d'Errico, R. Yates et al., "Emergence of Modern Human Behavior: Middle Stone Age Engravings from South Africa," *Science* 295 (2002): 1278–1280;

M. Balter, "Early Start for Human Art? Ochre May Revise Timeline," *Science* 323 (2009): 569; Stringer, *Lone Survivors*, 157.

6. R. Kittler, M. Kayser, and M. Stoneking, "Molecular Evolution of *Pediculus humanus* and the Origin of Clothing," *Current Biology* 13 (2003): 1414–1417; "Is This a Man?," *Economist*, December 24, 2005, 7; J. Travis, "The Naked Truth? Lice Hint at a Recent Origin of Clothing," *Science News Online* 164 (2003): 118, www.sciencenews.org/articles/20030823/fob7.asp.
7. Carl Zimmer, *Evolution: The Triumph of an Idea* (New York: HarperCollins, 2001), 305.
8. C. Zimmer, "How We Got Here: DNA Points to a Single Migration from Africa," *New York Times*, September 22, 2016; Brian Fagan, *People of the Earth: An Introduction to World Prehistory* (Upper Saddle River, NJ: Prentice Hall, 2004), 104; A. Lawler, "Did Modern Humans Travel out of Africa Via Arabia?," *Science* 331 (2011): 387.
9. A. Gibbons, "A New View of the Birth of *Homo sapiens*," *Science* 331 (2011): 392–394.
10. G. Hadjashov, T. Kivisild, P. A. Underhill et al., "Revealing the Prehistoric Settlement of Australia by Y Chromosome and mtDNA Analysis," *Proceedings of the National Academy of Sciences USA* 104 (2007): 8726–8730; N. Wade, "From DNA Analysis, Clues to a Single Australian Migration," *New York Times*, May 8, 2007; Robert J. Wenke and Deborah I. Olszewski, *Patterns in Prehistory: Mankind's First Three Million Years* (Oxford: Oxford University Press, 2007), 178. C. Gosden, "When Humans Arrived in the New Guinea Highlands," *Science* 330 (2010): 41–42; Andrew Shryock and Daniel L. Smail, *Deep History* (Berkeley: University of California Press, 2011), 203.
11. A. Gibbons, "Oldest *Homo sapiens* Genome Pinpoints Neandertal Input," *Science* 343 (2014): 1417; M. V. Anikovich, A. A. Sinitsyn, and J. F. Hoffecker, "Early Upper Paleolithic in Eastern Europe and Implications for the Dispersal of Modern Humans," *Science* 315 (2007): 223–225; "Modern Humans' First European Tour," *Science* 334 (2011): 576.
12. Zimmer, *Evolution*, 297; M. Balter, "Mild Climate, Lack of Moderns Let Last Neandertals Linger in Gibraltar," *Science* 313 (2006): 1557; P. Mellars and J. C. French, "Tenfold Population Increase in Western Europe at the Neandertal-to-Modern Human Transition," *Science* 333 (2011): 623–627; Steven Mithen, *The Prehistory of the Mind: The Cognitive Origins of Art, Religion and Science* (London: Thames and Hudson, 1996), 203, quoting Andrew Whiten.
13. H. Wimmer and J. Perner, "Beliefs About Beliefs: Representation and Constraining Function of Wrong Beliefs in Young Children's Understanding of Deception," *Cognition* 13 (1983): 103–128.
14. J. Perner and H. Wimmer, "'John *Thinks* That Mary *Thinks* That . . .': Attribution of Second-Order Beliefs by 5- to 10-Year-Old Children," *Journal of Experimental Child Psychology* 39 (1985): 437–471.
15. Nicholas Humphrey, *The Inner Eye: Social Intelligence in Evolution* (New York: Oxford University Press, 2002), 70–71; Zygmunt Bauman, *Mortality, Immortality and Other Life Strategies* (Stanford: Stanford University Press, 1992), 12.

16. Theodosius Dobzhansky, *The Biology of Ultimate Concern* (New York: New American Library, 1967), 52, 68; John C. Eccles, *Evolution of the Brain* (New York: Routledge, 1989), 236; Pierre Teilhard de Chardin, *The Phenomenon of Man* (New York: Harper and Row, 1965), 165, 180. Teilhard de Chardin took part in the excavation of *Homo erectus* fossils in China and was thus in a unique position to appreciate the implications of the emerging findings for Christian theology. He wrote *The Phenomenon of Man* in 1938, but the Catholic Church would not give him permission to publish it until 1955.
17. L. C. Aiello and R. I. M. Dunbar, "Neocortex Size, Group Size, and the Evolution of Language," *Current Anthropology* 34 (1993): 184–193; Robin Dunbar, *The Human Story: A New History of Mankind's Evolution* (London: Faber and Faber, 2004), 114–115, 125.
18. T. J. Crow, "Introduction," in Crow, *The Speciation of Modern* Homo sapiens, 7–8, quoting Bickerton; Terrence C. Deacon, *The Symbolic Species: The Co-Evolution of Language and the Brain* (New York: Norton, 1997); P. T. Schoenemann, "Evolution of the Size and Functional Areas of the Human Brain," *Annual Review of Anthropology* 35 (2006).
19. Steven Pinker, *How the Mind Works* (New York: Norton, 1997), 15, 362; Richard Passingham, *What Is Special About the Human Brain?* (Oxford: Oxford University Press, 2008), 9; Perner and Wimmer, "John *Thinks*."
20. Simon Baron-Cohen, *Mindblindness: An Essay on Autism and Theory of Mind* (Cambridge: MIT Press, 1997), 131; Mark Leary, *The Curse of Self: Self-Awareness, Egotism, and the Quality of Human Life* (Oxford: Oxford University Press, 2004), 390.
21. Pullum is quoted in P. Raffaele, "Speaking Bonobo," *Smithsonian*, November 2006, 74. Pinker is quoted in Michael R. Trimble, *The Soul in the Brain: The Cerebral Basis of Language, Art and Belief* (Baltimore: Johns Hopkins University Press, 2007), 57. George Washington Carver is quoted in P. V. Tobias, "Recent Advances in the Evolution of the Hominids with Special Reference to Brain and Speech," in *Recent Advances in the Evolution of Primates*, ed. Carlos Chagas (Vatican City: Pontificiae Academiae Scientiarum Scripta Varia 50, 1983), 85–140.
22. Deacon, *The Symbolic Species*, 281–292; Gerhard Roth, *The Long Evolution of Brains and Minds* (New York: Springer, 2013), 257.
23. Q. D. Atkinson, "Phonemic Diversity Supports a Serial Founder Effect Model of Language Expansion from Africa," *Science* 332 (2011): 346–349.
24. Zimmer, *Evolution*, 291, quoting Dunbar.
25. Mithen, *The Prehistory of the Mind*, 185.
26. de Chardin, *The Phenomenon of Man*, 165.
27. D. Bickerton, "From Protolanguage to Language," in *The Speciation of Modern* Homo sapiens, ed. Tim J. Crow (Oxford: Oxford University Press, 2002), 108.
28. L. van der Meer, S. Costafreda, A. Aleman et al., "Self-Reflection and the Brain: A Theoretical Review and Meta-Analysis of Neuroimaging Studies with Implications for Schizophrenia," *Neuroscience and Biobehavioral Reviews* 34 (2010): 935–946.
29. D. T. Stuss, "Disturbance of Self-Awareness After Frontal System Damage," in *Awareness of Deficit After Brain Injury: Clinical and Theoretical Issues*, ed. George P. Prigatano

and Daniel L. Schacter (New York: Oxford University Press, 1991), 68; D. M. Amodio and C. D. Frith, "Meeting of Minds: The Medial Frontal Cortex and Social Cognition," *Nature Reviews: Neuroscience* 7 (2006): 268–277.

30. G. Northoff and F. Bermpohl, "Cortical Midline Structures and the Self," *Trends in Cognitive Sciences* 8 (2004): 102–107; K. Tsapkini, C. E. Frangakis, and A. E. Hillis, "The Function of the Left Anterior Temporal Pole: Evidence from Acute and Stroke Infarct Volume," *Brain* 134 (2011): 3094–3105. The anterior temporal pole is also an area that is frequently damaged by repetitive trauma in contact sports such as football; see K. Willeumier, D. V. Taylor, and D. G. Amen, "Elevated Body Mass in National Football League Players Linked to Cognitive Impairment and Decreased Prefrontal Cortex and Temporal Pole Activity," *Translational Psychiatry* 2 (2012): e68; I. R. Olson, A. Plotzker, and Y. Ezzyat, "The Enigmatic Temporal Pole: A Review of Findings on Social and Emotional Processing," *Brain* 130 (2007): 1718–1731.

5. MODERN *HOMO SAPIENS*

1. P. Villa, S. Soriano, T. Tsanova et al., "Border Cave and the Beginning of the Later Stone Age in South Africa," *Proceedings of the National Academy of Sciences USA* 109 (2012): 13208–13213; R. Dale Guthrie, *The Nature of Paleolithic Art* (Chicago: University of Chicago Press, 2005), 29; C. Desdemaines-Hugon, *Stepping Stones: A Journey Through the Ice Age Caves of the Dordogne* (New Haven: Yale University Press, 2010), 75.
2. Chris Stringer, *Lone Survivors: How We Came to Be the Only Humans on Earth* (New York: Times, 2012), 150; Brian Fagan, *Cro-Magnon: How the Ice Age Gave Birth to the First Modern Humans* (New York: Bloomsbury, 2010), 167; See also M. Balter, "Clothes Make the (Hu) Man," *Science* 325 (2009): 1329.
3. David Lewis-Williams, *The Mind in the Cave* (London: Thames and Hudson, 2002), 221–222; S. A. de Beaune and R. White, "Ice Age Lamps," *Scientific American*, March 1993, 108–113.
4. S. O'Connor, R. Ono, and C. Clarkson, "Pelagic Fishing at 42,000 Years Before the Present and the Maritime Skills of Modern Humans," *Science* 334 (2011): 1117–1121.
5. Gregory Cochran and Henry Harpending, *The 10,000 Year Explosion: How Civilization Accelerated Human Evolution* (New York: Basic, 2009), 30; Steven Mithen, *The Prehistory of the Mind: The Cognitive Origins of Art, Religion and Science* (London: Thames and Hudson, 1996), 169.
6. Alexander Marshack, *The Roots of Civilization*, rev. ed. (1972; Mount Kisco, NY: Moyer Bell, 1991), 79; John C. Eccles, *Evolution of the Brain* (New York: Routledge, 1989), 135–136; Frederick L. Coolidge and Thomas Wynn, *The Rise of* Homo sapiens*: The Evolution of Modern Thinking* (New York: Wiley Blackwell, 2009), 234–235.
7. Lewis-Williams, *The Mind in the Cave*, 78; R. White, "Toward a Contextual Understanding of the Earliest Body Ornaments," in *The Emergence of Modern Humans*, ed. Erik Trinkaus (New York: Cambridge University Press, 1989), 211–231, at 213, 225–226; R. White, "Rediscovering French Ice-Age Art," *Nature* 320 (1986): 683–684.

8. S. McBrearty and A. S. Brooks, "The Revolution That Wasn't: A New Interpretation of the Origin of Modern Human Behavior," *Journal of Human Evolution* 39 (2000): 453–563. See also R. White, "Beyond Art: Toward an Understanding of the Origins of Material Representation in Europe," *Annual Review of Anthropology* 21 (1992): 537–564; R. White, "Technological and Social Dimensions of 'Aurignacian-Age' Body Ornaments Across Europe," in *Before Lascaux: The Complex Record of the Early Upper Paleolithic*, ed. Heidi Knecht, Anne Pike-Tay, and Randall White (Ann Arbor: CRC, 1992), 277–299; S. L. Kuhn, M. C. Stiner, D. S. Reese et al., "Ornaments of the Earliest Upper Paleolithic: New Insights from the Levant," *Proceedings of the National Academy of Sciences USA* 98 (2001): 7641–7646.
9. D. L. Smail and A. Shryock, "History and the 'Pre,'" *American Historical Review* 118 (2013): 709–737.
10. Ian Tattersall, *Becoming Human: Evolution and Human Uniqueness* (New York: Harcourt Brace, 1998), 162; Steve Olson, *Mapping Human History: Genes, Race, and Our Common Origins* (Boston: Houghton Mifflin, 2002), 73–76.
11. Tattersall, *Becoming Human*, 10.
12. Robin Dunbar, *The Human Story: A New History of Mankind's Evolution* (London: Faber and Faber, 2004), 187; B. Klima, "A Triple Burial from the Upper Paleolithic of Dolní Věstonice, Czechoslovakia," *Journal of Human Evolution* 16 (1988): 831–835; Brian Fagan, *People of the Earth: An Introduction to World Prehistory* (Upper Saddle River, NJ: Prentice Hall, 2004), 134; Guthrie, *The Nature of Paleolithic Art*, 142; Cochran and Harpending, *The 10,000 Year Explosion*.
13. T. Einwogerer, H. Friesinger, M. Handel et al., "Upper Palaeolithic Infant Burials," *Nature* 444 (2006): 285; Desdemaines-Hugon, *Stepping Stones*, 87; N. Wade, "24,000-Year-Old Body Shows Kinship to Europeans and American Indians," *New York Times*, November 21, 2013.
14. F. B. Harrold, "A Comparative Analysis of Eurasian Palaeolithic Burials," *World Archaeology* 12 (1980): 195–211.
15. Richard G. Klein and Blake Edgar, *The Dawn of Human Culture: A Bold New Theory on What Sparked the "Big Bang" of Human Consciousness* (New York: Wiley, 2002), 247–251. See also P. B. Beaumont, H. de Villiers, and J. C. Vogel, "Modern Man in Sub-Saharan Africa Prior to 49 000 Years B.P.: A Review and Evaluation with Particular Reference to Border Cave," *South African Journal of Science* 74 (1978): 409–419; A. Sillen and A. Morris, "Diagenesis of Bone from Border Cave: Implications for the Age of the Border Cave Hominids," *Journal of Human Evolution* 31 (1996): 499–506; J. Parkington, "A Critique of the Consensus View on the Age of Howieson's Poort Assemblages in South Africa," in *The Emergence of Modern Humans: An Archaeological Perspective*, ed. Paul Mellars (Ithaca: Cornell University Press, 1990), 34–55; Harrold, "A Comparative Analysis"; in support of Harrold's position, see also B. Hayden, "The Cultural Capacities of Neandertals: A Review and Re-Evaluation," *Journal of Human Evolution* 24 (1993): 113–146; Tattersall, *Becoming Human*, 162; in support of Tattersall's position, see also Christopher Stringer and Clive Gamble, *In Search of the Neanderthals* (London: Thames and Hudson, 1993), 158–161; Mithen, *The*

Prehistory of the Mind, 135–136; and M. Balter, "Did Neandertals Truly Bury Their Dead?," *Science* 337 (2012): 1443–1444.

16. Tattersall, *Becoming Human*, 161; Klein and Edgar, *The Dawn of Human Culture*, 192–193.
17. Jean Clottes and David Lewis-Williams, *The Shamans of Prehistory: Trance and Magic in the Painted Caves* (New York: Abrams, 1998), 114.
18. Klein and Edgar, *The Dawn of Human Culture*, 196, quoting Mellars; Fagan, *Cro-Magnon*, 234.
19. Gregory Curtis, *The Cave Painters: Probing the Mysteries of the World's First Artists* (New York: Anchor, 2006), 96; Claire Golomb, *Child Art in Context: A Cultural and Comparative Perspective* (Washington, DC: American Psychological Association, 2002), 100.
20. M. Aubert, A. Brumm, M. Ramli et al., "Pleistocene Cave Art from Sulawesi, Indonesia (Letter)," *Nature* 514 (2014): 223–227; J. Marchant, "The Awakening," *Smithsonian*, January–February 2016, 80–95; A. W. G. Pike, D. L. Hoffmann, M. Garciá-Diez et al., "U-Series Dating of Paleolithic Art in 11 Caves in Spain," *Science* 336 (2012): 1409–1413; David S. Whitley, *Cave Paintings and the Human Spirit: The Origin of Creativity and Belief* (Amherst, NY: Prometheus, 2009), 53; Evan Hadingham, *Secrets of the Ice Age: The World of the Cave Artists* (New York: Walker, 1979), 260–271.
21. C. Walker, "First Artists," *National Geographic*, January 2015, 33–57; N. J. Conard, "A Female Figurine from the Basal Aurignacian of Hohle Fels Cave in Southwestern Germany (Letter)," *Nature* 459 (2009): 248–252; Dunbar, *The Human Story*, 6; E. Culotta, "On the Origin of Religion," *Science* 326 (2009): 784–787, quoting Mellars; J. N. Wilford, "Flute's Revised Age Dates the Sound of Music Earlier," *New York Times*, May 29, 2012; M. Balter, "Early Dates for Artistic Europeans," *Science* 336 (2012): 1086–1087; Stringer, *Lone Survivors*, 122.
22. Fagan, *People of the Earth*, 129. See also Lyn Wadley, "The Pleistocene Later Stone Age South of the Limpopo River," *Journal of World Prehistory* 7 (1993): 243–296; D. Bruce Dickson, *The Dawn of Belief* (Tucson: University of Arizona Press, 1990); Paul G. Bahn, "New Advances in the Field of Ice Age Art," in *Origins of Anatomically Modern Humans*, ed. M. H. Nitecki and D. V. Nitecki (New York: Plenum, 1994), 121–132.
23. Clottes and Lewis-Williams, *The Shamans of Prehistory*, 115; J. Clottes, "Thematic Changes in Upper Paleolithic Art: A View from the Grotte Chauvet," *Antiquity* 70 (1996): 276–288; J. Clottes, "The 'Three Cs': Fresh Avenues Toward European Paleolithic Art," in *The Archaeology of Rock-Art*, ed. Christopher Chippindale and Paul S. C. Taçon (Cambridge: Cambridge University Press, 1998), 114; Curtis, *The Cave Painters*, 17.
24. Golomb, *Child Art in Context*, 106; M. Pruvost, R. Bellone, N. Benecke et al., "Genotypes of Predomestic Horses Match Phenotypes Painted in Paleolithic Works of Cave Art," *Proceedings of the National Academy of Sciences USA* 108 (2011): 18626–18630; W. Hunt, "Cave Painters Had a Leg up on Modern Painters," *Discover*, December 2013, 18.
25. Jean-Marie Chauvet, Eliette Brunel Deschamps, and Christian Hillaire, *Dawn of Art: The Chauvet Cave* (London: Thames and Hudson, 1996); John Pfeiffer, *The Creative*

Explosion: An Inquiry Into the Origins of Art and Religion (New York: Harper and Row, 1982), 1, 146; K. Turner, "Art with a Dark Past," *Washington Post*, July 30, 2000; J.-P. Rigaud, "Lascaux Cave: Art Treasures from the Ice Age," *National Geographic*, October 1988, 499; Andrew Shryock and Daniel L. Smail, *Deep History* (Berkeley: University of California Press, 2011), 131.

26. Curtis, *The Cave Painters*, 96, 114.
27. Brian Hayden, *Shamans, Sorcerers, and Saints* (Washington, DC: Smithsonian, 2003), 136; Curtis, *The Cave Painters*, 183–184.
28. Whitley, *Cave Paintings and the Human* Spirit, 65; M. Balter, "New Light on the Oldest Art," *Science* 283 (1999): 920–922; L.-H. Fage, "Hands Across Time: Exploring the Rock Art of Borneo," *National Geographic*, August 2005, 32–43; Paul Bahn, *Prehistoric Art* (Cambridge: Cambridge University Press, 1998), 112–115; M. Jenkins, "Last of the Cave People," *National Geographic*, February 2012, 127–141.
29. Clottes and Lewis-Williams, *The Shamans of Prehistory*, 46.
30. White, "Beyond Art," 558.
31. S. McBrearty and A. S. Brooks, "The Revolution That Wasn't: A New Interpretation of the Origin of Modern Human Behavior," *Journal of Human Evolution* 39 (2000): 453–563.
32. P. Schilder and D. Wechsler, "The Attitudes of Children Toward Death," *Journal of Genetic Psychology* 45 (1934): 406–451; D. J. Povinelli, K. R. Landau, and H. K. Perilloux, "Self-Recognition in Young Children Using Delayed Versus Live Feedback: Evidence of a Developmental Asynchrony," *Child Development* 67 (1996): 1540–1554. See also K. Nelson, "The Psychological and Social Origins of Autobiographical Memory," *Psychological Science* 4 (1993): 7–14; D. J. Povinelli, "The Unduplicated Self," in *The Self in Infancy: Theory and Research*, ed. P. Rochat (New York: Elsevier, 1995), 161–192; William James, *The Principles of Psychology* (1890; New York: Dover, 1950), 335.
33. J. S. DeLoache and N. M. Burns, "Early Understanding of the Representational Function of Pictures," *Cognition* 52 (1994): 83–110; J. DeLoache, "Mindful of Symbols," *Scientific American* 293 (2005): 72–77.
34. Gerhard Roth, *The Long Evolution of Brains and Minds* (New York: Springer, 2013), 11; C. M. Atance and D. K. O'Neill, "The Emergence of Episodic Future Thinking in Humans," *Learning and Motivation* 36 (2005): 126–144. Serious research on autobiographical memory dates to the work of Canadian neuroscientist Endel Tulving in the 1970s; see, for example, E. Tulving, "Episodic Memory: From Mind to Brain," *Annual Review of Psychology* 53 (2002): 1–25; Marcel Proust, *Swann's Way*, vol. 1, *Remembrance of Things Past*, trans. C. K. Scott-Moncrieff, Project Gutenberg, www.gutenberg.org/etext/7178.
35. Atance and O'Neill, "The Emergence of Episodic Future Thinking"; T. Suddendorf, "Episodic Memory Versus Episodic Foresight: Similarities and Differences," *WIREs Cognitive Science* 1 (2010): 99–107; J. Busby and T. Suddendorf, "Recalling Yesterday and Predicting Tomorrow," *Cognitive Development* 20 (2005): 362–372; T. Suddendorf, "Linking Yesterday and Tomorrow: Preschoolers' Ability to Report Temporally Displaced Events," *British Journal of Developmental Psychology* 28 (2010): 491–498;

Eccles, *Evolution of the Brain*, 229; T. Suddendorf, D. R. Addis, and M. C. Corballis, "Mental Time Travel and the Shaping of the Human Mind," *Philosophical Transactions of the Royal Society B* 364 (2009): 1317–1324. Suddendorf noted that research on the importance of foresight and its relationship to memory was singled out as one of the most significant breakthroughs by *Science* magazine in 2007 (see Suddendorf, "Episodic Memory").

36. T. S. Eliot, *The Complete Poems and Plays, 1909–1950* (New York: Harcourt, Brace, 1952); Lewis Carroll, *Alice's Adventures in Wonderland* and *Through the Looking Glass* (New York: Airmont, 1965), 181–182.
37. D. R. Addis, D. C. Sacchetti, B. A. Ally et al., "Episodic Stimulation of Future Events Is Impaired in Mild Alzheimer's Disease," *Neuropsychologia* 47 (2009): 2660–2671; Carl Zimmer, "The Brain," *Discover*, April 2011, 24–26; S. B. Klein and J. Loftus, "Memory and Temporal Experience: The Effects of Episodic Memory Loss on the Amnesic Patient's Ability to Remember the Past and Imagine the Future," *Social Cognition* 20 (2002): 353–379. See also Thomas Suddendorf, *The Gap: The Science of What Separates Us from Other Animals* (New York: Basic, 2013), 91, for a useful discussion of this.
38. For a summary of the lively debate about this question, see W. A. Roberts, "Mental Time Travel: Animals Anticipate the Future," *Current Biology* 17 (2007): R418–R420; N. S. Clayton, T. J. Bussey, and A. Dickenson, "Can Animals Recall the Past and Plan for the Future?," *Nature Reviews: Neuroscience* 4 (2003): 685–691; T. Suddendorf and M. C. Corballis, "Behavioural Evidence for Mental Time Travel in Nonhuman Animals," *Behavioural Brain Research* 215 (2010): 292–298; M. Balter, "Can Animals Envision the Future? Scientists Spar Over New Data," *Science* 340 (2013): 909.
39. Mithen, *The Prehistory of the Mind*, 168; Lewis-Williams, *The Mind in the Cave*, 78.
40. Lewis-Williams, *The Mind in the Cave*, 79; M. W. Conkey, "The Identification of Prehistoric Hunter-Gatherer Aggregation Sites: The Case of Altamira," *Current Anthropology* 21 (1980): 609–620.
41. Suddendorf, Addis, and Corballis, "Mental Time Travel"; Suddendorf, "Episodic Memory."
42. Edward B. Tylor, *Primitive Culture: Researches Into the Development of Mythology, Philosophy, Religion, Language, Art and Custom*, 2 vols. (1871; New York: Holt, 1874). Tylor cited Darwin's findings in vol. 2, pp. 152 and 223.
43. Mary Roach, *Stiff: The Curious Lives of Human Cadavers* (New York: Norton, 2003), 68, 70; Karina Croucher, *Death and Dying in the Neolithic Near East* (New York: Oxford University Press, 2012), 306; Raymond Tallis, *The Kingdom of Infinite Space* (New Haven: Yale University Press, 2008), 249.
44. William Shakespeare, *Hamlet*, act 5, scene 1; Theodosius Dobzhansky, *The Biology of Ultimate Concern* (New York: New American Library, 1967), 69.
45. Mike Parker Pearson, *The Archeology of Death and Burial* (College Station: Texas A and M University Press, 1999), 145. Tillich is quoted by Matthew Alper in *The "God" Part of the Brain* (New York: Rogue, 2001), 96. Baudelaire's *Les Fleurs Du Mal* is

quoted by Bauman in Zygmunt Bauman, *Mortality, Immortality and Other Life Strategies* (Stanford: Stanford University Press, 1992), 20. Vladimir Nabokov, *Speak, Memory* (New York: Vintage, 1989), 19. The T. S. Eliot line is from "The Waste Land" in *The Complete Poems and Plays*.

46. Daniel L. Pals, *Seven Theories of Religion* (New York: Oxford University Press, 1996), 24–25.
47. Nabakov, *Speak, Memory*, 77; M. H. Nagy, "The Child's View of Death," in *The Meaning of Death*, ed. Herman Feifel (New York: McGraw-Hill, 1959), 79–98. See also D. Y. Poltorak and J. P. Glazer, "The Development of Children's Understanding of Death: Cognitive and Psychodynamic Considerations," *Child and Adolescent Psychiatric Clinics of North America* 15 (2006): 567–573.
48. Nagy, "The Child's View of Death."
49. See, for example, Cynthia Moss, *Elephant Memories: Thirteen Years in the Life of an Elephant Family* (New York: Fawcett Columbine, 1988), 270–271; and D. Joubert, "Eyewitness to an Elephant Wake," *National Geographic*, May 1991, 39–41.
50. See G. Teleki, "Group Response to the Accidental Death of a Chimpanzee in Gombe National Park, Tanzania," *Folia Primatologica* 20 (1973): 81–94; and Jane Goodall, *The Chimpanzees of Gombe: Patterns of Behavior* (Cambridge: Harvard University Press, 1986), 330 (see also 109, 283–285); Edgar Morin, cited by Bauman, *Mortality, Immortality and Other Life Strategies*, 13.
51. Poems, in *Petronius, with an English translation by Michael Heseltine, and Seneca Apocolocynto, with an English translation by William Henry Denham Rouse*, 1913, 343, http://books.google.com/books?id=9DNJAAAAIAAJ&printsec=frontcover&dq=petroniu; Thomas Hobbes, *Leviathan*, chapter 12, 1651, Project Gutenberg EBook, www.gutenberg.org/ebooks/3207. See also Annemarie de Waal Malefijt, *Religion and Culture: An Introduction to Anthropology of Religion* (New York: Macmillan, 1968), 27–28; Erich Fromm, *The Anatomy of Human Destructiveness*, 302, quoted by Bauman, *Mortality, Immortality and Other Life Strategies*, 22; William Butler Yeats, "Death," in *Selected Poetry* (London: Pan, 1974), 142.
52. Ernest Becker, *The Denial of Death* (New York: Free, 1973), ix; P. T. P. Wong and A. Tomer, "Beyond Terror and Denial: The Positive Psychology of Death Acceptance (Editorial)," *Death Studies* 35 (2011): 99–106.
53. B. L. Burke, A. Martens, and E. H. Faucher, "Two Decades of Terror Management Theory: A Meta-Analysis of Mortality Salience Research," *Personality and Social Psychology Review* 14 (2010): 155–195; A. Rosenblatt, J. Greenberg, S. Solomon et al., "Evidence for Terror Management Theory: I. The Effects of Mortality Salience on Reactions to Those Who Violate or Uphold Cultural Values," *Journal of Personality and Social Psychology* 57 (1989): 681–690.
54. Tylor, *Primitive Culture*, 2:1.
55. A. Irving Hallowell, "The Role of Dreams in Ojibwa Culture," in *The Dream and Human Societies*, ed. G. E. Van Gruenbaum and Roger Caillois (Berkeley: University of California Press, 1966), 269.
56. Tylor, *Primitive Culture*, 1:441–443, 2:2.

57. Patrick McNamara, *The Neuroscience of Religious Experience* (Cambridge: Cambridge University Press, 2009), 203; Patrick McNamara and Kelly Bulkeley, "Dreams as a Source of Supernatural Agent Concepts," *Frontiers in Psychology* 6 (2015): 1–8.
58. Elizabeth Colson, *The Makah Indians: A Study of an Indian Tribe in Modern American Society* (Minneapolis: University of Minnesota Press, 1953), http://ehrafworldcultures.yale.edu/document?id=ne11–002; Alfred Métraux, *Myths and Tales of the Matako Indians (The Gran Chaco, Argentina)* (Gothenburg, Sweden: Walter Kaudern, 1939), http://ehrafworldcultures.yale.edu/document?id=si07–003.
59. Effie Bendann, *Death Customs: An Analytic Study of Burial Rites* (New York: Holt, 1930), 171, 257.
60. Clottes, in Christopher Chippendale and Paul S. C. Taçon, *The Archaeology of Rock Art* (Cambridge: Cambridge University Press, 1998), 125.
61. Curtis, *The Cave Painters*, 21; E. O. Wilson, "On the Origins of the Arts," *Harvard Magazine*, May–June 2012.
62. Curtis, *The Cave Painters*, 47.
63. Ibid., 210–211.
64. E. Fuller Torrey, *The Mind Game: Witchdoctors and Psychiatrists* (New York: Emerson Hall, 1972), 4–6.
65. Clottes and Lewis-Williams, *The Shamans of Prehistory*, 99; Lewis-Williams, *The Mind in the Cave*, 220; Whitley, *Cave Paintings and the Human Spirit*, 41–42; Hayden, *Shamans, Sorcerers, and Saints*, 142. The picture drawn by Breuil of "the sorcerer," which is included in virtually every textbook of human development and archeology, is in fact much more impressive than pictures of the original cave drawing on which it was based, leading some observers to conclude that Breuil was exaggerating some features. See Pfeiffer, *The Creative Explosion*, 108.
66. É. Durkheim, "The Elementary Forms of Religious Life," in *A Reader in the Anthropology of Religion*, ed. Michael Lambek (Malden, Mass.: Blackwell, 2002), 46; William A. Lessa and Evon Z. Vogt, *Reader in Comparative Religion*, 4th ed. (New York: Harper and Row, 1979), 27, 9; William James, *The Varieties of Religious Experience* (1902; New York: Random House, 1929), 31–34.
67. Curtis, *The Cave Painters*, 209, 195, 99, 142–144; Whitley, *Cave Paintings and the Human Spirit*, 32–33.
68. Clottes and Lewis-Wilson, *The Shamans of Prehistory*, 69–71.
69. Guthrie, *The Nature of Paleolithic Art*, 9–10.
70. Tylor, *Primitive Culture*, 1:483.
71. Tattersall, *Becoming Human*, 10; Mithen, *The Prehistory of the Mind*, 175–176; Tylor, *Primitive Culture*, 1:486.
72. R. N. Spreng, R. A. Mar, and S. N. Kim, "The Common Neural Basis of Autobiographical Memory, Prospection, Navigation, Theory of Mind, and the Default Mode: A Quantitative Meta-Analysis," *Journal of Cognitive Neuroscience* 21 (2009): 489–510; J. S. Rabin, A. Gilboa, D. T. Stuss et al., "Common and Unique Neural Correlates of Autobiographical Memory and Theory of Mind," *Journal of Cognitive Neuroscience* 22 (2010): 1095–1111; H. C. Lou, B. Luber, M. Crupain et al., "Parietal Cortex and

Representation of the Mental Self," *Proceedings of the National Academy of Sciences USA* 101 (2004): 6827–6832.

73. P. Pioline, G. Chételat, V. Matuszewski et al., "In Search of Autobiographical Memories: A PET Study in the Frontal Variant of Frontotemporal Dementia," *Neuropsychologia* 45 (2007): 2730–2743; S. Oddo, S. Lux, P. H. Weiss et al., "Specific Role of Medial Prefrontal Cortex in Retrieving Recent Autobiographical Memories: An fMRI Study of Young Female Subjects," *Cortex* 46 (2010): 29–39; D. Stuss and B. Levine, "Adult Clinical Neuropsychology: Lessons from Studies of the Frontal Lobes," *Annual Review of Psychology* 53 (2002): 401–433; D. T. Stuss, "Disturbance of Self-Awareness After Frontal System Damage," in *Awareness of Deficit After Brain Injury: Clinical and Theoretical Issues*, ed. George P. Prigatano and Daniel L. Schacter (New York: Oxford University Press, 1991).

74. J. Okuda, T. Fujii, H. Ohtake et al., "Thinking of the Future and Past: The Roles of the Frontal Pole and the Medial Temporal Lobes," *NeuroImage* 19 (2003): 1369–1380; D. R. Addis, A. T. Wong, and D. L. Schacter, "Remembering the Past and Imagining the Future: Common and Distinct Neural Substrates During Event Construction and Elaboration," *Neuropsychologia* 45 (2007): 1363–1377; D. L. Schacter and D. R. Addis, "The Ghosts of the Past and Future," *Nature* 445 (2007): 27.

75. C. Lebel, L. Walker, A. Leemans et al., "Microstructural Maturation of the Human Brain from Childhood to Adulthood," *NeuroImage* 40 (2008).

76. N. C. Andreasen, D. S. O'Leary, S. Paradiso et al., "The Cerebellum Plays a Role in Conscious Episodic Memory Retrieval," *Human Brain Mapping* 8 (1999): 226–234; G. R. Fink, H. J. Markowitsch, M. Reinkemeier et al., "Cerebral Representation of One's Own Past: Neural Networks Involved in Autobiographical Memory," *Journal of Neuroscience* 16 (1996): 4275–4282. See also E. Svoboda, M. C. McKinnon, and B. Levine, "The Functional Neuroanatomy of Autobiographical Memory: A Meta-Analysis," *Neuropsychologia* 44 (2006): 2189–2208; Coolidge and Wynn, *The Rise of* Homo sapiens, 24; J. H. Balsters, E. Cussans, J. Diedrichsen et al., "Evolution of the Cerebellar Cortex: The Selective Expansion of Prefrontal-Projecting Cerebellar Lobules," *NeuroImage* 49 (2010): 2045–2052; A. H. Weaver, "Reciprocal Evolution of the Cerebellum and Neocortex in Fossil Humans," *Proceedings of the National Academy of Sciences USA* 102 (2005): 3576–3580.

77. Addis et al., "Remembering the Past"; Schacter and Addis, "The Ghosts of the Past."

6. ANCESTORS AND AGRICULTURE

1. P. Kareiva, S. Watts, R. McDonald et al., "Domesticated Nature: Shaping Landscapes and Ecosystems for Human Welfare," *Science* 316 (2007): 1866–1869.

2. R. Dale Guthrie, *The Nature of Paleolithic Art* (Chicago: University of Chicago Press, 2005), 406; W. Dansgaard, J. W. C. White, and S. J. Johnsen, "The Abrupt Termination of the Younger Dryas Climate Event," *Nature* 339 (1989): 532–534; Peter Bellwood, *First Farmers: The Origin of Agricultural Societies* (Malden, MA: Blackwell, 2005),

19–25. The millennium of colder weather is called the Younger Dryas, after an arctic flower. It is thought to have been caused by huge amounts of glacial water flooding the North Atlantic and causing major changes in weather patterns.

3. O. Dietrich, C. Köksal-Schmidt, J. Notroff et al., "First Came the Temple, Later the City," *Actual Archaeology Magazine*, Summer 2012, 32–51; Klaus Schmidt, *Göblecki Tepe: A Stone Age Sanctuary in South-Eastern Anatolia* (Munich: Beck, 2012).
4. A. Curry, "The World's First Temple?," *Smithsonian*, November 2008, 54–60; Patrick E. McGovern, *Uncorking the Past: The Quest for Wine, Beer, and Other Alcoholic Beverages* (Berkeley: University of California Press, 2009), 81.
5. A. Curry, "Seeking the Roots of Ritual," *Science* 319 (2008): 278–280; Curry, "The World's First Temple?"
6. M. Rosenberg, "Hallan Çemi," in *Neolithic in Turkey*, ed. M. Ozdoğan (Istanbul: Arkeoloji ve Sanat Yayinlari, 1999), 25–33.
7. Schmidt, *Göblecki Tepe*, 69–76; McGovern, *Uncorking the Past*, 77–78.
8. Schmidt, *Göblecki Tepe*, 57–58; Karina Croucher, *Death and Dying in the Neolithic Near East* (Oxford: Oxford University Press, 2012), 221.
9. Alan H. Simmons, *The Neolithic Revolution in the Near East: Transforming the Human Landscape* (Tucson: University of Arizona Press, 2007), 151.
10. Schmidt, *Göblecki Tepe*, 231; Croucher, *Death and Dying*, 134; C. C. Mann, "The Birth of Religions," *National Geographic*, June 2011, 39–59.
11. Croucher, *Death and Dying*, 139; Schmidt, *Göblecki Tepe*, 69; Dietrich et al., "First Came the Temple"; Curry, "The World's First Temple?"
12. Edward B. Tylor, *Primitive Culture: Researches Into the Development of Mythology, Philosophy, Religion, Language, Art and Custom*, 2 vols. (1871; New York: Holt, 1874), 1:427; James L. Cox, *The Invention of God in Indigenous Societies* (Durham: Acumen, 2014), 4; H. C. People, P. Duda, and F. W. Marlowe, "Hunter-Gatherers and the Origin of Religion," *Human Nature* 27 (2016): 261–282.
13. John Bailey, "Account of the Wild Tribes of the Veddahs of Ceylon: Their Habits, Customs, and Superstitions," in *Transactions*, vol. 2 (London: Ethnological Society of London, 1863), 301–302, http://ehrafworldcultures.yale.edu/document?id=ax05-002; C. G. Seligman, Brenda Z. Seligman, Charles Samuel Myers et al., Gunasekara, *The Veddas*, Cambridge Archaeological and Ethnological Series (Cambridge: Cambridge University Press, 1911), 30, http://ehrafworldcultures.yale.edu/document?id=ax05-001; Allan R. Holmberg, *Nomads of the Long Bow: The Siriono of Eastern Bolivia*, Smithsonian Institution, Institute of Social Anthropology (Washington, DC: Government Printing Office, 1950), 89, http://ehrafworldcultures.yale.edu/document?id=sf21-001; Pew Forum on Religion and Public Life Survey, August 2009, question 292a (Pew Research Center for People and the Press, 2012), 54.
14. Charles A. Bishop, *The Northern Ojibwa and the Fur Trade: An Historical and Ecological Study*, Cultures and Communities, Native Peoples (Toronto: Holt, Rinehart and Winston of Canada, 1974), 7, http://ehrafworldcultures.yale.edu/document?id=ng06-054; A. Irving Hallowell and Jennifer S. H. Brown, *The Ojibwa of Berens River, Manitoba: Ethnography Into History*, Case Studies in Cultural Anthropology

(Fort Worth: Harcourt Brace Jovanovich, 1991), 76, http://ehrafworldcultures.yale.edu/document?id=ng06-058; Clark Wissler, *Societies and Dance Associations of the Blackfoot Indians*, Anthropological Papers of the American Museum of Natural History (New York: Trustees, 1913), 443, http://ehrafworldcultures.yale.edu/document?id=nf06-018; Kaj Birket-Smith, *The Chugach Eskimo*, Nationalmuseets Skrifter, Etnografisk Række (Kobenhavn: Nationalmuseets publikationsfond, 1953), 112–113, http://ehrafworldcultures.yale.edu/document?id=na10-001; Frederica De Laguna, *Under Mount Saint Elias: The History and Culture of the Yakutat Tlingit*, Smithsonian Contributions to Anthropology (Washington, DC: Smithsonian, 1972), 606, http://ehrafworldcultures.yale.edu/document?id=na12-020.

15. Geoffrey Parrinder, *African Traditional Religion* (London: Hutchinson University Library, 1954), 57–66; Lorna Marshall, "!Kung Bushman Religious Beliefs," in *Africa*, vol. 32 (London: Oxford University Press, 1962), 241, http://ehrafworldcultures.yale.edu/document?id=fx10-013; Lorna Marshall, *The !Kung of Nyae Nyae* (Cambridge: Harvard University Press, 1976), 53, http://ehrafworldcultures.yale.edu/document?id=fx10-017; Tylor, *Primitive Culture*, 1:422. See also L. B. Steadman, C. T. Palmer, and C. T. Tilley, "The Universality of Ancestor Worship," *Ethnology* 35 (1996): 63–76.
16. Jacques Cauvin, *The Birth of the Gods and the Origins of Agriculture* (Cambridge: Cambridge University Press, 2000), 11, originally published in 1994 as *Naissance des Divinities, Naissance de l'Agriculture* (Paris, CNRS); Jared Diamond, *Guns, Germs, and Steel: The Fates of Human Societies* (New York: Norton, 1997), 140.
17. Robert J. Wenke and Deborah I. Olszewski, *Patterns in Prehistory: Mankind's First Three Million Years* (Oxford: Oxford University Press, 2007), 250; Steven Mithen, *The Prehistory of the Mind: The Cognitive Origins of Art, Religion and Science* (London: Thames and Hudson, 1996), 218; M. Balter, "Seeking Agriculture's Ancient Roots," *Science* 316 (2007): 1830–1835, quoting Douglas Kennett at the University of Oregon.
18. G. Willcox, "The Roots of Civilization in Southwestern Asia," *Science* 341 (2013): 39–40; Wenke and Olszewski, *Patterns in Prehistory*, 251.
19. McGovern, *Uncorking the Past*, 82, 13.
20. B. Hayden, N. Canuel, and J. Shanse, "What Was Brewing in the Natufian? An Archaeological Assessment of Brewing Technology in the Epipaleolithic," *Journal of Archaeological Method and Theory* 20 (2013): 102–150.
21. Ibid.; McGovern, *Uncorking the Past*, xiii, 81.
22. Chris Stringer, *Lone Survivors: How We Came to Be the Only Humans on Earth* (New York: Times, 2012), 166; Guthrie, *The Nature of Paleolithic Art*, 407–408; E. Pennisi, "Old Dogs Teach a New Lesson About Canine Regions," *Science* 342 (2013): 785–786.
23. "Sheep Domestication Caught in the Act," *Science* 344 (2014): 456; Juliet Clutton-Brock, *Domesticated Animals from Early Times* (Austin: University of Texas Press, 1981), 57–58.
24. Croucher, *Death and Dying*, 3, 24. Archeologists divide the Neolithic era into periods based on the presence of pottery and other items: Natufian (14,500–12,000 years ago); prepottery Neolithic A (12,000–10,500 years ago); early prepottery Neolithic B (10,500–10,100 years ago); middle prepottery Neolithic B (10,100–9,300 years ago);

late prepottery Neolithic B (9,300–8,700 years ago); final prepottery Neolithic B (8,700–8,300 years ago); and pottery Neolithic (8,300–7,200 years ago).

25. P. Skoglund, H. Malmström, M. Raghavan et al., "Origins and Genetic Legacy of Neolithic Farmers and Hunter-Gatherers in Europe," *Science* 336 (2012): 466–469; R. Bouckaert, P. Lemey, M. Dunn et al., "Mapping the Origins and Expansion of the Indo-European Language Family," *Science* 337 (2012): 957–960; Greger Larson, "How Wheat Came to Britain," *Science* 347 (2015): 945–946.
26. M. Balter, "New Light on Revolutions That Weren't," *Science* 336 (2012): 530–531; Wenke and Olszewski, *Patterns in Prehistory*, 375.
27. Gregory Cochran and Henry Harpending, *The 10,000 Year Explosion: How Civilization Accelerated Human Evolution* (New York: Basic, 2009), 31; Wenke and Olszewski, *Patterns in Prehistory*, 230.
28. X. Wu, C. Zhang, P. Goldberg et al., "Early Pottery at 20,000 Years Ago in Xianrendong Cave, China," *Science* 336 (2012): 1696–1700; G. Shelach, "On the Invention of Pottery," *Science* 336 (2012): 1644–1645; Wenke and Olszewski, *Patterns in Prehistory*, 261; Andrew Shryock and Daniel L. Smail, *Deep History* (Berkeley: University of California Press, 2011), 211; McGovern, *Uncorking the Past*, 39; A. Tucker, "Dig, Drink and Be Merry," *Smithsonian*, July–August 2011, 38–48.
29. Bellwood, *First Farmers*, 141–145.
30. Richard L. Burger, *Chavin and the Origins of Andean Civilization* (London: Thames and Hudson, 1992), 42; Wenke and Olszewski, *Patterns in Prehistory* 538–539, 262–268; Bellwood, *First Farmers*, 106–110.
31. Croucher, *Death and Dying*, 303; M. J. Rossano, "Supernaturalizing Social Life," *Human Nature* 18 (2007): 272–294; Brian Hayden, *Shamans, Sorcerers, and Saints* (Washington, DC: Smithsonian, 2003), 184–185; Mike Parker Pearson, *The Archaeology of Death and Burial* (College Station: Texas A and M University Press, 1999), 161.
32. Croucher, *Death and Dying*, 56, 238, 290.
33. Ibid., 243.
34. Steve Olson, *Mapping Human History: Genes, Race, and Our Common Origins* (Boston: Houghton Mifflin, 2002), 97; Parker Pearson, *The Archaeology of Death and Burial*, 158.
35. Croucher, *Death and Dying*, 36, 41, 213.
36. Cauvin, *The Birth of the Gods*, 81.
37. Croucher, *Death and Dying*, 94–95; Cauvin, *The Birth of the Gods*, 113. It should be noted that plastered skulls were not unique to people in southwest Asia during the agricultural revolution and have occasionally been found elsewhere in the world. For example, in early 2012, a plastered skull from early twentieth-century Papua New Guinea was on display at the Musee du Quai Branly in Paris.
38. Jacquetta Hawkes, *The Atlas of Early Man* (New York: St. Martin's, 1976), 41; Michael Balter, *The Goddess and the Bull: Çatalhöyük: An Archeological Journey to the Dawn of Civilization* (Walnut Creek, CA: Left Coast, 2006), 282, quoting Kathleen Kenyon, *Digging Up Jericho*; Croucher, *Death and Dying*, 152–153.
39. Croucher, *Death and Dying*, 143, 145, 214; Parker Pearson, *The Archaeology of Death and Burial*, 159.

40. Parker Pearson, *The Archaeology of Death and Burial*, 161; Schmidt, *Göblecki Tepe*, 38; Croucher, *Death and Dying*, 45, 124; "The Nahal Hemar Mask," *Current World Archeology* 66 (2014): 66; H.-D. Bienert, "The Er-Ram Stone Mask at the Palestine Exploration Fund, London," *Oxford Journal of Archaeology* 9 (1990): 257–261.
41. Croucher, *Death and Dying*, 150.
42. Ibid., 47; Simmons, *The Neolithic Revolution in the Near East*, 154–155.
43. Hawkes, *The Atlas of Early Man*, 41; Balter, *The Goddess and the Bull*, 42.
44. M. Balter, "The Seeds of Civilization," *Smithsonian*, May 2005, 68–74; Hawkes, *The Atlas of Early Man*, 41–42.
45. Croucher, *Death and Dying*, 111, 188; Wenke and Olszewski, *Patterns in Prehistory* 332–333; Balter, *The Goddess and the Bull*, 30, 37; Cauvin, *The Birth of the Gods*, 31; Hawkes, *The Atlas of Early Man*, 41.
46. Wenke and Olszewski, *Patterns in Prehistory*, 333; Croucher, *Death and Dying*, 111.
47. McGovern, *Uncorking the Past*, 33.
48. Ibid., 40–41.
49. Michael E. Moseley, *The Incas and Their Ancestors: The Archaeology of Peru* (New York: Thames and Hudson, 1992), 86–87; T. D. Dillehay, J. Rossen, T. C. Andres et al., "Preceramic Adoption of Peanut, Squash, and Cotton in Northern Peru," *Science* 316 (2007): 1890–1893; Bellwood, *First Farmers*, 99; O. Hanotte, D. G. Bradley, J. W. Ochieng et al., "African Pastoralism: Genetic Imprints of Origins and Migrations," *Science* 296 (2002): 336–339; P. C. Sereno, E. A. A. Garcea, H. Jousse et al., "Lakeside Cemeteries in the Sahara: 5000 Years of Holocene Population and Environmental Change," *PLoS ONE* 3 (2008): 1–22; Salima Ikram, *Death and Burial in Ancient Egypt* (London: Longman, 2003), 23; Kenneth L. Feder, *The Past in Perspective: An Introduction to Human History* (Mountain View, CA: Mayfield, 2000), 406–407.
50. Annemarie deWaal Malefijt, *Religion and Culture: An Introduction to Anthropology of Religion* (New York: Macmillan, 1968), 18–19; Tylor, *Primitive Culture*, 2:247.
51. Herbert Basedow, *The Australian Aboriginal* (Adelaide: F. W. Preece and Sons, 1925), http://ehrafworldcultures.yale.edu/document?id=oi08-007.
52. Edward L. Schieffelin and Robert Crittenden, *Like People You See in a Dream: First Contact in Six Papuan Societies* (Stanford: Stanford University Press, 1991), 74, 101, 171, 222. See also Bob Connolly and Robin Anderson, *First Contact: New Guinea's Highlanders Encounter the Outside World* (New York: Viking, 1987); and Edward Marriott, *The Lost Tribe: A Harrowing Passage Into New Guinea's Heart of Darkness* (New York: Holt, 1996).
53. Schieffelin and Crittenden, *Like People You See*, 63, 92, 94; Croucher, *Death and Dying*, 125.
54. Cochran and Harpending, *The 10,000 Year Explosion*, 65; C. Haub, "How Many People Have Ever Lived on Earth?," Population Research Bureau, www.prb.org/Articles/2002/HowManyPeopleHaveEverLivedonEarth.aspx.
55. George P. Murdoch, *Ethnographic Atlas* (Pittsburgh: University of Pittsburgh Press, 1967), 52 (the data can be accessed online); Guy Swanson, *The Birth of the Gods* (Ann Arbor: University of Michigan Press, 1960), 42, 56; F. L. Roes and M. Raymond, "Belief in Moralizing Gods," *Evolution and Human Behavior* 24 (2003): 126–135;

A. F. Shariff, "Big Gods Were Made for Big Groups," *Religion, Brain and Behavior* 1 (2011): 89–93.

56. Cauvin, *The Birth of the Gods*, 112.
57. Parker Pearson, *The Archaeology of Death and Burial*, 164. See also Simmons, *The Neolithic Revolution in the Near East*, 157.
58. Cauvin, *The Birth of the Gods*, 112; Balter, *The Goddess and the Bull*, x, 37–39; I. Hodder, "Women and Men at Çatalhöyük," *Scientific American* 290 (2004): 76–83; Cauvin, *The Birth of the Gods*, 32.
59. Balter, *The Goddess and the Bull*, 322.
60. "Ancestor Worship," *Encyclopedia Britannica* (Chicago: Encyclopedia Britannica, 1954), 1:888.
61. Guthrie, *The Nature of Paleolithic Art*, 405.
62. Georg F. Striedter, *Principles of Brain Evolution* (Sunderland, MA: Sinauer Associates, 2005), 333.
63. Paul E. Flechsig, *Anatomie des menschlichen Gehirns und Rückenmarks auf myelogenetischer Grundlage* (Leipzig: Thieme, 1920); N. Gogtay, J. N. Giedd, L. Lusk et al., "Dynamic Mapping of Human Cortical Development During Childhood Through Early Adulthood," *Proceedings of the National Academy of Sciences USA* 101 (2004): 8174–8179; J. N. Giedd, "Structural Magnetic Resonance Imaging of the Adolescent Brain," *Annals of the New York Academy of Sciences* 1021 (2004): 77–85; T. M. Preuss, "Evolutionary Specializations of Primate Brain Systems," in *Primate Origins: Adaptations and Evolution*, ed. Matthew J. Ravosa and Marian Dagasto (New York: Springer, 2007), 625–675; John Allman, *Evolving Brains* (New York: Scientific American Library, 2000), 176; T. M. Preuss, "Primate Brain Evolution in Phylogenetic Context," in *Evolution of Nervous Systems*, vol. 4, *Primates*, ed. Jon H. Kaas and Todd M. Preuss (Oxford: Elsevier, 2007), 1–34.
64. P. T. Schoenemann, M. J. Sheehan, and L. D. Glotzer, "Prefrontal White Matter Volume Is Disproportionately Larger in Humans Than in Other Primates," *Nature Neuroscience* 8 (2005): 242–225.

7. GOVERNMENTS AND GODS

1. Peter Bellwood, *First Farmers: The Origin of Agricultural Societies* (Malden, MA: Blackwell, 2005), 15; J. Nicholas Postgate, *Early Mesopotamia: Society and Economy at the Dawn of History* (London: Routledge, 1992), 112.
2. Postgate, *Early Mesopotamia*, 206–221; Samuel N. Kramer, *The Sumerians: Their History, Culture, and Character* (Chicago: University of Chicago Press, 1963), 73–111.
3. Brian Fagan, *People of the Earth: An Introduction to World Prehistory* (Upper Saddle River, NJ: Prentice Hall, 2004), 362–363; Kramer, *The Sumerians*, 73, 135.
4. Kramer, *The Sumerians*, 73–74.
5. Thorkild Jacobsen, *The Treasures of Darkness: A History of Mesopotamian Religion* (New Haven: Yale University Press, 1976), 26.
6. Ibid., 110–111.

7. Ibid., 20, 36.
8. Ibid., 27; Kramer, *The Sumerians*, 110–111; Patrick E. McGovern, *Uncorking the Past: The Quest for Wine, Beer, and Other Alcoholic Beverages* (Berkeley: University of California Press, 2009), 98; Glyn Edmund Daniel, *The First Civilizations: The Archaeology of Their Origins* (New York: Crowell, 1968), 74. Other English words said to have been derived from Mesopotamian, specifically Sumerian, language origins include *gypsum*, *myrrh*, *saffron*, and *naphtha*.
9. Jacobsen, *The Treasures of Darkness*, 47, 36.
10. Kramer, *The Sumerians*, 132, 134, 154; George Roux, *Ancient Iraq*, 3rd ed. (London: George Allen and Unwin, 1964; New York: Penguin, 1992), 100; Julian Jaynes, *The Origins of Consciousness in the Breakdown of the Bicameral Mind* (Boston: Houghton Mifflin, 1976), 162.
11. Kramer, *The Sumerians*, 126; M. Dirda, "In Search of Gilgamesh, the Epic Hero of Ancient Babylonia," *Washington Post Book World*, March 4, 2007; N. K. Sandars, *The Epic of Gilgamesh*, rev. ed. (New York: Penguin, 1972), 101–102. Gilgamesh is thought to have been a real ruler who ruled in Sumer about 4,700 years ago.
12. Sandars, *The Epic of Gilgamesh*, 102, 106, 107, 115, 119.
13. Jacobsen, *The Treasures of Darkness*, 20, 73.
14. Ibid., 83.
15. Roux, *Ancient Iraq*, 169; R. L. Zettler, "The Royal Cemetery of Ur," in *Treasures from the Royal Tombs of Ur*, ed. Richard L. Zettler and Lee Horne (Philadelphia: University of Pennsylvania Museum, 1998), 21–32, at 25; D. P. Hansen, "Art of the Royal Tombs of Ur: A Brief Interpretation," in Zettler and Horne, *Treasures*, 47.
16. R. L. Zettler, "The Burials of a King and Queen," in Zettler and Horne, *Treasures*, 35–36; Roux, *Ancient Iraq*, 137.
17. Postgate, *Early Mesopotamia*, 109, 118, 120; Kramer, *The Sumerians*, 117–118, 123; Roux, *Ancient Iraq*, 99.
18. Kramer, *The Sumarians*, 136–137; Postgate, *Early Mesopotamia*, 114–115, 135–136.
19. Postgate, *Early Mesopotamia*, 126–127; Kramer, *The Sumarians*, 142; Roux, *Ancient Iraq*, 132.
20. Jacobsen, *The Treasures of Darkness*, 78; Postgate, *Early Mesopotamia*, 252.
21. Roux, *Ancient Iraq*, 138–139, 141–142.
22. Postgate, *Early Mesopotamia*, 133, 253; Kramer, *The Sumarians*, 261, 90.
23. Roux, *Ancient Iraq*, 23. Xenophanes is cited by Clyde Kluckhohn, "Foreword," in *Reader in Comparative Religion: An Anthropological Approach*, ed. William A. Lessa and Evon Z. Vogt (New York: Harper and Row, 1979), v–vi; Baron de La Brède Montesquieu, *Lettres Persones* (Paris: Alphonse Lemerre, 1721), 59.
24. Roux, *Ancient Iraq*, 85.
25. Robert J. Wenke and Deborah I. Olszewski, *Patterns in Prehistory: Mankind's First Three Million Years* (Oxford: Oxford University Press, 2007), 382–383.
26. Ibid., 389–390; Edith Hamilton, *The Greek Way to Western Civilization* (New York: Norton, 1930), 13; Salima Ikram, *Death and Burial in Ancient Egypt* (London: Longman, 2003), ix.
27. Ikram, *Death and Burial*, 152.

28. Kenneth L. Feder, *The Past in Perspective: An Introduction to Human History* (Mountain View, CA: Mayfield, 2000), 402; Ikram, *Death and Burial*, 152–153.
29. A complete description of the mummification process can be found in Ikram, *Death and Burial*, and in Carol Andrews, *Egyptian Mummies* (Cambridge: Harvard University Press, 1984).
30. Andrews, *Egyptian Mummies*, 83; Ikram, *Death and Burial*, 81–82.
31. Andrews, *Egyptian Mummies*, 30, 72.
32. Ikram, *Death and Burial*, 132, 200; McGovern, *Uncorking the Past*, 167.
33. Ikram, *Death and Burial*, 128–131; Andrews, *Egyptian Mummies*, 75, 79.
34. Bruce G. Trigger, *Understanding Early Civilizations* (New York: Cambridge University Press, 2003), 409.
35. Feder, *The Past in Perspective*, 409–410; Wenke and Olszewski, *Patterns in Prehistory*, 417; A. Lawler, "The Indus Script—Write or Wrong?," *Science* 306 (2004): 2026–2029. See also Burjor Avari, *India: The Ancient Past* (New York: Routledge, 2007), 44–45.
36. Bridget Allchin and Raymond Allchin, *The Rise of Civilization in India and Pakistan* (Cambridge: Cambridge University Press, 1982), 213; Mortimer Wheeler, *The Indus Civilization* (Cambridge: Cambridge University Press, 1962), 89; Avari, *India*, 48.
37. Allchin and Allchin, *The Rise of Civilization*, 217, 238, 305; A. Lawler, "Boring No More, a Trade-Savvy Indus Emerges," *Science* 320 (2008): 1276–1281.
38. David W. Anthony, ed., *The Lost World of Old Europe: The Danube Valley, 5000–3500 BC* (Princeton: Princeton University Press, 2010), 29.
39. Marija Gimbutas, *The Gods and Goddesses of Old Europe: Myths and Cult Images* (Berkeley: University of California Press, 1982), 11, 195; Douglas W. Bailey, "The Figurines of Old Europe," in Anthony, *The Lost World*, 113–127.
40. The finds at Varna are described by C. Renfrew, "Varna and the Social Context of Early Metallurgy," *Antiquity* 52 (1978): 199–203; C. Renfrew, "Varna and the Emergence of Wealth in Prehistoric Europe," in *The Social Life of Things: Commodities in Cultural Perspective*, ed. Arjun Appadurai (Cambridge: Cambridge University Press, 1986), 141–168; Mike Parker Pearson, *The Archaeology of Death and Burial* (College Station: Texas A and M University Press, 1999), 79; and J. N. Wilford, "A Lost European Culture, Pulled from Obscurity," *New York Times*, December 1, 2009.
41. Renfrew, "Varna and the Emergence of Wealth."
42. C. Desdemaines-Hugon, *Stepping Stones: A Journey Through the Ice Age Caves of the Dordogne* (New Haven: Yale University Press, 2010), 144–145; J. O'Shea and M. Zvelebil, "Oleneostrovski Mogilnik: Reconstructing the Social and Economic Organization of Prehistoric Foragers in Northern Russia," *Journal of Anthropological Archaeology* 3 (1984): 1–40.
43. M. J. O'Kelly, "The Megalithic Tombs of Ireland," in *The Megalithic Monuments of Western Europe*, ed. Colin Renfrew (London: Thames and Hudson, 1983), 113–126, at 113; R. Chapman, "The Emergence of Formal Disposal Areas and the 'Problem' of Megalithic Tombs in Prehistoric Europe," in *The Archeology of Death*, ed. Robert Chapman, Ian Kinnes, and Klaves Randborg (Cambridge: Cambridge University Press, 1981), 71.

44. Jean-Pierre Mohen, *Standing Stones: Stonehenge, Carnac, and the World of Megaliths* (London: Thames and Hudson, 1999), 82–83.
45. Ibid., 55.
46. C. Renfrew, "Introduction: The Megalithic Builders of Western Europe," in Renfrew, *The Megalithic Monuments*, 8–17, 9; Mohen, *Standing Stones*, 57; P.-R. Giot, "The Megaliths of France," in Renfrew, *The Megalithic Monuments*, 18–28, 26–27. See also B. Bramanti, M. G. Thomas, W. Haak et al., "Genetic Discontinuity Between Local Hunter-Gatherers and Central Europe's First Farmers," *Science* 326 (2009): 137–140.
47. Wenke and Olszewski, *Patterns in Prehistory*, 462.
48. M. Balter, "Monumental Roots," *Science* 343 (2014): 18–23; Roff Smith, "Before Stonehenge," *National Geographic*, August 2014, 26–51.
49. Caroline Malone, *The Prehistoric Monuments of Avebury* (Swindon: National Trust, 1994), 38, 39, 47.
50. Ibid., 21–25.
51. Ibid., 10–13.
52. Mark Gillings and Joshua Pollard, *Avebury* (London: Gerald Duckworth, 2004), 72–73.
53. Aubrey Burl, *A Guide to the Stone Circles of Britain, Ireland and Brittany* (New Haven: Yale University Press, 1995), 87.
54. Timothy Darvill, *Long Barrows of the Cotswolds and Surrounding Areas* (Stroud, Gloucestershire: Tempus, 2004), 165–168, 212; Malone, *The Prehistoric Monuments*, 29–32; Brian Hayden, *Shamans, Sorcerers, and Saints* (Washington, DC: Smithsonian, 2003), 229.
55. Aubrey Burl, *Prehistoric Stone Circles* (Aylesbury: Shire, 1979), 10, 42; Gillings and Pollard, *Avebury*, 63–64.
56. Kwang-chih Chang, *The Archeology of Ancient China*, 4th ed. (New Haven: Yale University Press, 1986), 248; Wenke and Olszewski, *Patterns in Prehistory*, 432.
57. Robert H. Bellah, *Religion in Human Evolution: From the Paleolithic to the Axial Age* (Cambridge: Harvard University Press, 2011), 250–251.
58. A. Lawler, "Beyond the Yellow River: How China Became China," *Science* 325 (2009): 930–935; Trigger, *Understanding Early Civilizations*, 422.
59. Feder, *The Past in Perspective*, 412; Chang, *The Archeology of Ancient China*, 255, 276; McGovern, *Uncorking the Past*, 51.
60. R. S. Solis, J. Haas, and W. Creamer, "Dating Caral, a Preceramic Site in the Supe Valley on the Central Coast of Peru," *Science* 292 (2001): 723–726.
61. P. J. McDonnell, "Plaza in Peru May Be the Americas' Oldest Urban Site," *Los Angeles Times*, February 26, 2008; Richard L. Burger, *Chavin and the Origins of Andean Civilization* (London: Thames and Hudson, 1992), 80.
62. Burger, *Chavin*, 35–36.
63. Solis et al., "Dating Caral."
64. Ruth Shady Solis, Marco Machacuay Romero, Daniel Caceda Guillén et al., *Caral, the Oldest Civilization in the Americas: 15 Years Unveiling Its History* (Lima: Institute Nacional de Cultura, 2009), 46–53. See also Solis et al., "Dating Caral"; C. C. Mann,

"Oldest Civilization in the Americas Revealed," *Science* 307 (2005): 34–35; J. Haas and A. Ruiz, "Power and the Emergence of Complex Polities in the Peruvian Preceramic," *Archaeological Papers of the American Anthropological Association* 14 (2005): 37–52; J. Haas and W. Creamer, "Crucible of Andean Civilization: The Peruvian Coast from 3000 to 1800 BC," *Current Anthropology* 47 (2006): 745–775.

65. J. A. Lobell, "Atacama's Decaying Mummies," *Archaeology*, September–October 2015; Michael E. Moseley, *The Incas and Their Ancestors: The Archaeology of Peru* (New York: Thames and Hudson, 1992), 93–94, 144; Fagan, *People of the Earth*, 527.
66. H. Hoag, "Oldest Evidence of Andean Religion Found," *Nature*, April 15, 2003, www.nature.com/news/2003/030415/full/news030414-4.html; Mann, "Oldest Civilization."
67. Émile Durkheim, *The Elementary Forms of Religious Life* (1912; Oxford: Oxford University Press, 2001), 314; Theodosius Dobzhansky, *The Biology of Ultimate Concern* (New York: New American Library, 1967), 94, quoting Toynbee.
68. Burger, *Chavin*, 175, 149–150. See also Feder, *The Past in Perspective*, 378.
69. Moseley, *The Incas and Their Ancestors*, 155.
70. Karen Armstrong, *The Great Transformation: The Beginning of Our Religious Traditions* (New York: Knopf, 2006), 390.
71. Karl Jaspers, *The Future of Mankind* (Chicago: University of Chicago Press, 1961), 135; John Hick, *An Interpretation of Religion: Human Responses to the Transcendent* (New Haven: Yale University Press, 2004), 31; E. Weil, "What Is a Breakthrough in History?," *Daedalus* 104 (1975): 21–36; Karen Armstrong, *A History of God: The 4,000-Year Quest of Judaism, Christianity, and Islam* (New York: Ballantine, 1993), 27.
72. The inscription on Babylon's Royal Way can be seen in the Pergamon Museum in Berlin. William James, *The Varieties of Religious Experience* (New York: Random House, 1929), 514, first published in 1902. Luther is quoted in Corliss Lamont, *The Illusions of Immortality* (1935; New York: Continuum, 1990), 2.
73. H. Horn, "Where Does Religion Come From?," *Atlantic*, August 17, 2011, www.theatlantic.com/entertainment/archive/2011/08/where-does-religion-come- from/243723/.
74. Bismarck is quoted in Diamond, *Guns, Germs, and Steel*, 420.
75. Annemarie deWaal Malefijt, *Religion and Culture: An Introduction to Anthropology of Religion* (New York: Macmillan, 1968), 17. See also Arthur Cotterell and Rachel Storm, *The Ultimate Encyclopedia of Mythology* (London: Hermes House, 1999), 21; and Armstrong, *The Great Transformation*, 106.
76. Kramer, *The Sumerians*, 292–296; Armstrong, *A History of God*, 23; for the influence of the Persians on the Judaeans, see Isaiah 45:1 and Ezra 1:2 and 6:3–8; Mary Boyce, *Zoroastrians: Their Religious Beliefs and Practices* (Boston: Routledge and Kegan Paul, 1979), 51–53, 76–77, 99, 152–153. According to Zoroastrian theology, a virgin birth would be possible because it was said that Zoroaster's semen had been preserved in a lake, "and in the course of time each of the three virgins . . . will bathe there and conceive a son by the prophet, and each of these three sons will have his share in furthering the work of redemption." See Peter Clark, *Zoroastrianism: An Introduction to an Ancient Faith* (Portland, OR: Sussex Academic Press, 1998), 65–67; Mary Boyce, *A History of Zoroastrianism*, vol. 1 (New York: Brill, 1989), 285; Boyce, *Zoroastrians*,

154–155; and Richard Foltz, *Spirituality in the Land of the Noble: How Iran Shaped the World's Religions* (London: Oneworld, 2004), 25.

77. Robin Dunbar, *The Human Story: A New History of Mankind's Evolution* (London: Faber and Faber, 2004), 183, 197; Armstrong, *A History of God*, xix, 4, 362.
78. Ernest Becker, *The Denial of Death* (New York: Free, 1973), 26, 51.

8. OTHER THEORIES OF THE ORIGINS OF GODS

1. John Micklethwait and Adrian Wooldridge, *God Is Back: How the Global Revival of Faith Is Changing the World* (New York: Penguin, 2009), 134.
2. Charles Darwin, *The Descent of Man, and Selection in Relation to Sex* (London: John Murray, 1871), pt. 2, pp. 67, 68.
3. Sam Harris, *The End of Faith* (New York: Norton, 2004), 38.
4. Edward B. Tylor, *Primitive Culture: Researches Into the Development of Mythology, Philosophy, Religion, Language, Art and Custom*, 2 vols. (1871; New York: Holt, 1874), 2:2.
5. Daniel L. Pals, *Seven Theories of Religion* (New York: Oxford University Press, 1996), 114, 112, 89; Émile Durkheim, *The Elementary Forms of Religious Life* (1912; Oxford: Oxford University Press, 2001), 46.
6. Nicholas Wade, *The Faith Instinct: How Religion Evolved and Why It Endures* (New York: Penguin, 2009), 58, 10, 2, 9; Barbara J. King, *Evolving God: A Provocative View on the Origins of Religion* (New York: Doubleday, 2007), 7, 56.
7. David Sloan Wilson, *Darwin's Cathedral: Evolution, Religion, and the Nature of Society* (Chicago: University of Chicago Press, 2002), 165.
8. Wade, *The Faith Instinct*, 280.
9. Pascal Boyer, *Religion Explained: The Evolutionary Origins of Religious Thought* (New York: Basic, 2001), 23; M. Bateson, D. Nettle, and G. Roberts, "Cues of Being Watched Enhance Cooperation in Real-World Setting," *Biology Letters* 2 (2006): 412–414. This theme is also well summarized in Dominic Johnson and Jesse Bering, "Hand of God, Mind of Man: Punishment and Cognition in the Evolution of Cooperation," in *The Believing Primate: Scientific, Philosophical, and Theological Reflections on the Origin of Religion*, ed. Jeffrey Schloss and Michael I. Murray (New York: Oxford University Press, 2009), 26–43. For the study of children, see S. Vogt, C. Efferson, J. Berger et al., "Eye Spots Do Not Increase Altruism in Children," *Evolution and Human Behavior*, 2015, doi:10.1016/j.evolhumbehav.2014.11.007. One of the strengths of prosocial theories are attempts by its adherents to subject it to scientific verification; see, for example, B. G. Purzycki, C. Apicella, Q. D. Atkinson et al., "Moralistic Gods, Supernatural Punishment and the Expansion of Human Sociality (Letter)," *Nature* 530 (2016): 327–330.
10. Jesse Bering, *The Faith Instinct* (New York: Norton, 2011), 190.
11. Ara Norenzayan, *Big Gods: How Religion Transformed Cooperation and Conflict* (Princeton: Princeton University Press, 2013); Dominic Johnson, *God Is Watching You: How the Fear of God Makes Us Human* (New York: Oxford University Press,

2016). The two books are very similar, as is detailed by Johnson, "Big Gods, Small Wonder, Supernatural Punishment Strikes Back," *Religion, Brain and Behavior* 5 (2015): 290–298. For a good summary of prosocial theories, see also A. Norenzayan, A. F. Shariff, W. M. Gervais et al., "The Cultural Evolution of Prosocial Religions," *Behavioral and Brain Sciences*, 2016, doi:10.1017/S0140525X14001356.

12. Johnson, *God Is Watching You*, 3, 96, 73.
13. Karen Armstrong, *A History of God: The 4,000-Year Quest of Judaism, Christianity, and Islam* (New York: Ballantine, 1993), 389; Robert H. Bellah, *Religion in Human Evolution: From the Paleolithic to the Axial Age* (Cambridge: Harvard University Press, 2011), 1.
14. Patrick McNamara, *The Neuroscience of Religious Experience* (Cambridge: Cambridge University Press, 2009), 41, 163, 258.
15. Pals, *Seven Theories of Religion*, 79.
16. Coke Newell, *Latter Days: An Insider's Guide to Mormonism, the Church of Jesus Christ of Latter-Day Saints* (New York: St. Martin's Griffin, 2000), 240, 236, 241–242.
17. Robert A. Hinde, *Why Gods Persist: A Scientific Approach to Religion* (London: Routledge, 1999), 67; David J. Linden, *The Accidental Mind: How Brain Evolution Has Given Us Love, Memory, Dreams, and God* (Cambridge: Harvard University Press, 2007), 225.
18. Lionel Tiger and Michael McGuire, *God's Brain* (Amherst, NY: Prometheus, 2010), 20, 202–204.
19. Boyer, *Religion Explained*, 21; Stewart Guthrie, *Faces in the Clouds: A New Theory of Religion* (New York: Oxford University Press, 1993), 13.
20. Hinde, *Why Gods Persist*, 215, 216; Theodosius Dobzhansky, *The Biology of Ultimate Concern* (New York: New American Library, 1967), 92.
21. Stewart Guthrie, *Faces in the Clouds*, 3, 7, 6.
22. Michael Shermer, *How We Believe: The Search for God in an Age of Science* (New York: Freeman, 2000), 38–39; Boyer, *Religion Explained*, 318, 330; Daniel C. Dennett, *Breaking the Spell: Religion as a Natural Phenomenon* (New York: Viking, 2006), 109, 114.
23. McNamara, *The Neuroscience of Religious Experience*. See especially chapter 5.
24. Matthew Alper, *The "God" Part of the Brain* (New York: Rogue, 2001), 113; V. S. Ramachandran and Sandra Blakeslee, *Phantom in the Brain: Probing the Mysteries of the Human Mind* (New York: HarperCollins, 1998), 179; Michael A. Persinger, *Neuropsychological Bases of God Beliefs* (New York: Praeger, 1987), 14, 19.
25. D. De Ridder, K. Van Laere, P. Dupont et al., "Visualizing out-of-Body Experience in the Brain," *New England Journal of Medicine* 357 (2007): 1829–1833; P. Brugger, M. Regard, and T. Landis, "Unilaterally Felt 'Presences': The Neuropsychiatry of One's Invisible *Doppelgänger*," *Neuropsychiatry, Neuropsychology, and Behavioral Neurology* 9 (1996): 114–122; C. Urgesi, S. M. Aglioti, M. Skrap et al., "The Spiritual Brain: Selective Cortical Lesions Modulate Human Self-Transcendence," *Neuron* 65 (2010): 309–319; Alper, *The "God" Part of the Brain*, 188.
26. R. Joseph, "The Limbic System and the Soul: Evolution and the Neuroanatomy of Religious Experience," *Zygon* 36 (2001): 105–136; A. D. Owen, R. D. Hayward, H. G.

Koenig et al., "Religious Factors and Hippocampal Atrophy in Late Life," *PLoS One* 6 (2011): e17006; McNamara, *The Neuroscience of Religious Experience*, xi, 245.

27. Andrew Newberg and Mark R. Waldman, *Why We Believe What We Believe* (New York: Free, 2006), 175–176. See also Eugene d'Aquili and Andrew G. Newberg, *The Mystical Mind: Probing the Biology of Religious Experience* (Minneapolis: Fortress, 1999); R. D. Hayward, A. D. Owen, H. G. Koenig et al., "Associations of Religious Behavior and Experiences with Extent of Regional Atrophy in the Orbitofrontal Cortex During Older Adulthood," *Religion, Brain and Behavior* 1 (2011): 103–118; M. Inzlicht, A. M. Tullett, and M. Good, "The Need to Believe: A Neuroscience Account of Religion as a Motivated Process," *Religion, Brain and Behavior* 1 (2011): 192–251; N. P. Azari, J. Nickel, G. Wunderlich et al., "Neural Correlates of Religious Experience," *European Journal of Neuroscience* 13 (2001): 1649–1652.
28. McNamara, *The Neuroscience of Religious Experience*, xi; D. Kapogiannis, A. K. Barbey, M. Su et al., "Neuroanatomical Variability in Religiosity," *PLoS ONE* 4 (2009): e7180.
29. R. Dale Guthrie, *The Nature of Paleolithic Art* (Chicago: University of Chicago Press, 2005), 440; Nancy L. Segal, *Born Together—Reared Apart: The Landmark Minnesota Twin Study* (Cambridge: Harvard University Press, 2012), 144, 252.
30. Alper, *The "God" Part of the Brain*, 78, 82.
31. Dean Hamer, *The God Gene: How Faith Is Hardwired Into Our Genes* (New York: Anchor, 2004), 9–12, 139.
32. Julian Jaynes, *The Origins of Consciousness in the Breakdown of the Bicameral Mind* (Boston: Houghton Mifflin, 1976), 143.
33. R. M. Henig, "God Has Always Been a Puzzle," *New York Times Magazine*, March 4, 2007, 37–85, at 39. For an extended discussion of this issue, see J. P. Schloss and M. J. Murray, "Evolutionary Accounts of Belief in Supernatural Punishment: A Critical Review," *Religion, Brain and Behavior* 1 (2011): 46–99, and the book edited by them: Schloss and Murray, *The Believing Primate*.
34. A. F. Shariff and A. Norenzayan, "God Is Watching You," *Psychological Science* 18 (2007): 803–809; Wade, *The Faith Instinct*, 9–10.
35. Hamer, *The God Gene*, 10; Alper, *The "God" Part of the Brain*, 102; McNamara, *The Neuroscience of Religious Experience*, 28.
36. C. S. Alcorta, "Religion, Health, and the Social Signaling Model of Religion," *Religion, Brain and Behavior* 1 (2012): 213–216.
37. Scott Atran, *In Gods We Trust: The Evolutionary Landscape of Religion* (New York: Oxford University Press, 2002), 279; Richard Dawkins, *The God Delusion* (Boston: Houghton Mifflin, 2006), 172.
38. This contest is described in the Old Testament, 1 Kings 18:20–40. It was also memorialized by Felix Mendelssohn in his oratorio *Elijah*.
39. Matthew White, *The Great Big Book of Horrible Things: The Definitive Chronicle of History's 100 Worst Atrocities* (New York: Norton, 2012), 107, 112; Sam Harris, *Letter to a Christian Nation* (New York: Knopf, 2006; New York: Vintage, 2008), xii, 91.
40. Dostoevsky is quoted by Dobzhansky, *The Biology of Ultimate Concern*, 63; J. Gorden Melton, ed., *The Encyclopedia of American Religions: Creeds* (Detroit: Gale Research,

1988); John Micklethwait and Adrian Wooldridge, *God Is Back: How the Global Revival of Faith Is Changing the World* (New York: Penguin, 2009), 215; James G. Frazer, *The Fear of the Dead in Primitive Religion* (New York: Collier-MacMillan, 1933; New York: Biblo and Tannen, 1966), vi.

41. Jesus, a Humble Prophet of God, Al Islam, www.alislam.org/topics/jesus/index.php; S. Aziz, "Death of Jesus," bulletin, October 2001, Ahmadiyya Anjuman Ishaat Islam Lahore, UK, www.aaiil.org/uk/newsletters/2001/1001ukbulletin.pdf; A. A. Chaudhry, "The Promised Mahdi and Messiah," Islam International Publications Limited, www.alislam.org/library/books/promisedmessiah/index.htm?page=50; James E. Talmage, *Jesus the Christ* (Salt Lake City: Church of Jesus Christ of Latter-Day Saints, 1981), 721–736.
42. Michael Balter, *The Goddess and the Bull: Çatalhöyük: An Archeological Journey to the Dawn of Civilization* (Walnut Creek, CA: Left Coast, 2006), 320–321.
43. Timothy Darvill, *Long Barrows of the Cotswolds and Surrounding Areas* (Stroud, Gloucestershire: Tempus, 2004), 239; Robert Silverberg, *The Mound Builders* (Athens: Ohio University Press, 1970), 204–205; Moundbuilders Country Club, "The Beginning," www.moundbuilderscc.com.
44. Percy Bysshe Shelley, "Ozymandias," 1818, www.rc.umd.edu/rchs/reader/ozymandias.html.

APPENDIX A: THE EVOLUTION OF THE BRAIN

1. Gerhard Roth, *The Long Evolution of Brains and Minds* (New York: Springer, 2013), 234.
2. Ibid., 235. The glial cells that make myelin are called oligodendrocytes.
3. P. T. Schoenemann, M. J. Sheehan, and L. D. Glotzer, "Prefrontal White Matter Volume Is Disproportionately Larger in Humans Than in Other Primates," *Nature Neuroscience* 8 (2005): 242–252; T. Sakai, A. Mikami, M. Tomonaga et al., "Differential Prefrontal White Matter Development in Chimpanzees and Humans," *Current Biology* 21 (2011): 1397–1402; David C. Geary, *The Origin of Mind: Evolution of Brain, Cognition, and General Intelligence* (Washington, DC: American Psychological Association, 2005), 230. See also J. K. Rilling and T. R. Insel, "The Primate Neocortex in Comparative Perspective Using Magnetic Resonance Imaging," *Journal of Human Evolution* 37 (1999): 191–233; and C. C. Sherwood, R. L. Holloway, K. Semendeferi et al., "Is Prefrontal White Matter Enlargement a Human Evolutionary Specialization? (Letter)," *Nature Neuroscience* 8 (2005): 537–538.
4. O. Langworthy, "Development of Behavior Patterns and Myelinization of the Nervous System in the Human Fetus and Infant," *Contributions to Embryology* 139 (1933): 1–57. Regarding this, see also P. I. Yakovlev and A.-R. Lecours, "The Myelogenetic Cycles of Regional Maturation of the Brain," in *Regional Development of the Brain in Early Life*, ed. Alexandre Minkowski (Oxford: Blackwell, 1967), 3–70, at 64–66.
5. Paul E. Flechsig, *Anatomie des menschlichen Gehirns und Rückenmarks auf myelogenetischer Grundlage* (Leipzig: Thieme, 1920). Flechsig's studies have been largely

replicated by others; see Percival Bailey and Gerhardt von Bonin, *The Isocortex of Man* (Urbana: University of Illinois Press, 1951), 265. For timing of myelination, see F. M. Benes, "Myelination of Cortical-Hippocampal Relays During Late Adolescence," *Schizophrenia Bulletin* 15 (1989): 585–593; and Yakovlev and Lecours, "The Myelogenetic Cycles," 61.

6. Rilling and Insel, "The Primate Neocortex"; K. Zilles, E. Armstrong, A. Schleicher et al., "The Human Pattern of Gyrification in the Cerebral Cortex," *Anatomy and Embryology* 179 (1988): 173–179; N. W. Ingalls, "The Parietal Region in the Primate Brain," *Journal of Comparative Neurology* 24 (1914): 291–341; Bailey and von Bonin, *The Isocortex of Man*, 49; R. Holloway, "Evolution of the Human Brain," in *Handbook of Human Symbolic Evolution*, ed. Andrew Lock and Charles R. Peters (Oxford: Clarendon, 1996), 74–125, at 83. There are other measures of brain evolution, such as the formation of nerve connections (synapses) that also point in this direction, but these have not been as well studied. See P. R. Huttenlocher, C. De Courten, L. J. Garey et al., "Synaptic Development in Human Cerebral Cortex," *International Journal of Neurology* 16–17 (1982–1983): 144–154; and P. R. Huttenlocher and A. S. Dabholkar, "Regional Differences in Synaptogenesis in Human Cerebral Cortex," *Journal of Comparative Neurology* 387 (1997): 167–178. Such studies have shown that the formation of synapses in the prefrontal cortex occur later than in other brain areas.

INDEX

Accidental Mind, The (Linden), 212
Adaptionists, 218–20
Adonis, 201
Africa, 69–71, 236; ancestor worship in, 139–40
Afterlife, 125, 152, 165, 211–13. *See also* Ancestor worship; Dreams
Afterworld, 169–70, 181–82
Age, 44–45; axial, 198–202. *See also* Children
Aggregation sites, 110, 134
Agriculture, 205; ancestor worship and, 147–49, 153–55; autobiographical memory without, 161–64; burial related to, 147, 149; climate and, 133–34; diffusion of, 144–45; domestication in, 20, 133, 140–44; as epiphenomena, 137; parallel evolution of, 145–46; villages and, 148
Ahmadiyya, 222
Ahman, Mirza Ghulam, 222
Ahura Mazda, 201, 240*n*3
Aiello, Leslie, 78
Ainu, 238
Akhenaten, 196
Alaska, 231–32
Alcohol, 141–42, 145–46, 154
Alper, Matthew, 219
Altamira cave art, 96–97, 99, 103, 110
Alzheimer's disease, 43
Americans, 1–2, 221
Amon, 177
Amsterdam, Beulah, 41
Amygdala, 24, 46–47, 216–17; autobiographical memory and, 126, 127–28
Ancestor worship, 4, 151, 191, 205; agriculture and, 147–49, 153–55; cave art and, 124; deification from, 156, 165; by hunter-gatherers, 138–40; in Papua New Guinea, 156–58; parallel evolution of, 20; places for, 135–38; power from, 155–56; reasons for, 138–39, 204; shrines and, 152–53; skull cults as, 149–53; Tylor on, 140, 156, 204; white people and, 157. *See also* Dreams
Ancient Rome, 2, 115–16
Andaman Islanders, 237
Andrews, Carol, 180
Animal domestication, 143–44
Animals, 100; autobiographical memory and, 108; in cave art, 98–99, 121, 124;

Animals (*cont.*)
migration of, 109; mummification of, 180; spirits as, 157; theory of mind and, 57–59; without introspective self, 77
Animism, 114, 234
Anterior cingulate (BA 24 and 32), 46–48, 47; intelligence and, 31–32, 32; theory of mind and, 60, 63, 64, 84, 85, 86, 126, 126
Anterior insula, 47–49, 63; introspective self and, 84, 85, 86
Anterior temporal pole, 256*n*30
Anthropomorphism, 83, 197, 213–14; of Mesopotamian gods, 172, 175–76
Aphrodite, 200
Archaic *Homo Sapiens,* 68. *See also* Neandertals
Archeology of Death and Burial, The (Pearson), 23, 147–48
Arcuate fasciculus, 14, 15, 62
Armstrong, Karen, 199, 202
Arts, 19, 71; children related to, 105; of modern *Homo sapiens,* 95–101, 102–3; by Neandertals, 95. *See also* Cave art
Aryans, 196
Asclepius, 160
Association areas, 32–33
Association cortex, 33, 62
Assyria, 196
Astronomy, 188
Atkinson, Quentin, 81
Atran, Scott, 220
Aurochs (cattle), 144
Australia, 18, 73, 119, 123
Australopithecus, 25–26, 242*n*6
Austria, 93, 97
Autism, 60; belief in God and, 66
Autobiographical memory (episodic memory), 204, 205, 259*n*34; animals and, 108; of children, 104–5; future related to, 106–8, 128–29; grave goods and, 124–25; of *Homo sapiens,* 3–4, 16, 19; human revolution and, 120–21; hunting with, 108–10; language and, 110; of modern *Homo sapiens,* 104–9, 125–29, 126; Neandertals and, 108–9; uncinate fasciculus and, 128; understanding of death and, 110–13; visual arts and, 120–22; without agriculture, 161–64
Avebury, xiii–xiv, 187–90
Axial age, 198–202

Ba, 179
BA 6. *See* Premotor cortex
BA 7. *See* Superior parietal
BA 8. *See* Frontal cortex
BA 9. *See* Lateral prefrontal cortex
BA 10. *See* Frontal pole
BA 22. *See* Posterior superior temporal area
BA 23. *See* Posterior cingulate
BA 24. *See* Anterior cingulate
BA 32. *See* Anterior cingulate
BA 35. *See* Parahippocampal gyrus
BA 36. *See* Parahippocampal gyrus
BA 38. *See* Temporal pole
BA 39. *See* Inferior parietal lobe
BA 40. *See* Inferior parietal lobe
BA 46. *See* Lateral prefrontal cortex
BA 47. *See* Orbital frontal cortex; Orbitofrontal region
Baboons, 24–25, 42, 58, 115
Babylon, 169, 196, 199, 201, 269*n*11, 272*n*72
Baldwin, James, 133
Balter, Michael, 36, 152
Barama River Caribs, 234
Baron-Cohen, Simon, 60, 80
Basedow, Herbert, 156
Batek, 237–38
Baudelaire, Charles, 113–14
Bauman, Zygmunt, 76
Becker, Ernest, 116, 202
Beer, 141–43, 168–69
Belief in Gods, 1–2, 65–67
Belief Instinct, The (Bering), 65
Bella Coola Indians, 233

Bellah, Robert, 200
Bellwood, Peter, 146
Bering, Jesse, 65
Bicameral mind, 218
Bickerton, Derek, 53, 79, 84
Birds, 27
Bismarck, Otto von, 200
Bison, 99–100
Blackfoot Indians, 139
Boars (pigs), 144
Bolivia, 119, 139, 235
Bone tools, 70, 88, 102
Bonobos, 28
Book of the Dead, The, 182
Bow and arrow, 89, 103
Boyer, Pascal, 212, 214, 220
Brain abnormalities, 43, 49–50, 127; semantic memory and, 107–8
Brain evolution theory, 225–26. *See also* specific topics
Brains: of children, 13–14; of chimpanzees, 12–13, 30, 33, 163, 226–27; mammalian, 18, 23–24; of monkeys, 18–19; postmortem, 12–13; of primates, 33. *See also* Autobiographical memory; Human brain
Brain size, 10, 12–13, 243*n*12; of Hominins, 25–26; of *Homo erectus,* 38, 40, 45–46; of *Homo habilis,* 26–27, 29–30, 37–38; of *Homo sapiens,* 29, 84; of Neandertals, 52, 68–69, 84
Brazil, 119, 235
Breaking the Spell (Dennett), 214
Breuil, Henri, 122–23, 262*n*65
British Guiana, 234
Brodgar, Scotland, 187–88
Brodmann, Korbinian, 7, 8, 240*n*9, 245*n*20
Brooks, Alison, 101
Buddhism, 196, 198–200
Burger, Richard, 197
Burial, xiv, 171–72; agriculture related to, 147, 149; beneath houses, 149, 152; in Egypt, 178–79; of Neandertals, 54–55, 92, 94–95; place for, 151–52; stones for, 186–87. *See also* Grave goods
Burial mounds, 223
Burl, Aubrey, 189–90
Buttermore, Nicole, 20

Cairns, 186–87
California, 231
Canada, 139, 232–34
Canela Indians, 235
Cannibalism, 115
Caral, Peru, 193
Carroll, Lewis, 107
Carved human figures, 134–36, 138
Carver, George Washington, 80
Çatalhöyük , Turkey, 152–53, 159–60
Cattle (aurochs), 144
Cauvin, Jacques, 149, 153, 159–60
Cave art, 96–97, 102–3, 110; animals in, 98–99, 121, 124; geometric figures in, 100–1; handprints in, 100, 121; human figures in, 99–100; religion related to, 121–24; shamans in, 122
Çayönü, Turkey, 136–37
Cereals, 140–42
Cerebellum, 126, 128
Ceremonial buildings, 135–37
Ceremonies, 135–36, 139
Ceylon, 138–39
Chauvet cave art, 96–98, 100, 102, 121–23
Children, 15–16; autobiographical memory of, 104–5; brains of, 13–14; lateral prefrontal cortex of, 163; self-awareness of, 40–41; temples and, 173; theory of mind and, 56–57, 74–75, 82; understanding of death related to, 114–15
Chimpanzee brains, 12–13, 30, 33, 163, 226–27
Chimpanzee-hominin split, 11, 25
Chimpanzees, 28, 108; monkeys compared to, 43–44, 49; parallel evolution and, 25; self-awareness of, 43–44; understanding of death related to, 115

China, xiii–xiv, 190, 197; alcohol in, 145–46, 154; grave goods in, 191–92
Chipewyan Indians, 234
Christianity, 77, 201, 234, 255*n*16
Cingulate. *See* Anterior cingulate; Posterior cingulate
Cingulum, 14, 15, 128
City-states, 174–75
Climate, 39–40, 161, 263*n*2; agriculture and, 133–34
Clothing, fitted, 71, 89
Clottes, Jean, 121
Clutton-Brock, Juliet, 144
Cognitive domains, 8–9
Collins, Francis, 2
Colorado, 230
Comanche Indians, 230
Comfort theories, 211–13
Communication, 139
Conscription, 35
Convergent evolution, 242*n*18
Cooking, 39
Cooperative hunting, 109–10, 134
Corpus callosum, 34
Cosquer cave art, 98, 123
Cox, James, 138
Craig, Bud, 41–42
Cree Indians, 232–33
Creek Indians, 229
Crete, 197
Crick, Francis, 1
Critchley, Macdonald, 6–7
Croucher, Karina, 137–38, 144, 147, 149–50
Czech Republic, 93

d'Aquili, Eugene, 216
Darwin, Charles, 4–6, 43, 203–4, 240*n*7; Tylor and, 110–11, 206
Darwin's Cathedral (Wilson), 208
Dating, 11, 88
Dawkins, Richard, 220
Dawn of Human Culture, The (Klein and Edgar), 95
Deacon, Terrence, 79
Death, 139, 195, 199, 202; decomposition after, 111–12; Egypt and, 178–82; fear of, 113–14; Hominins and, 111–12; human brain after, 9; land of, 119–20; Mesopotamian gods and, 169–70; in wars, 174–75. *See also* Ancestor worship; Burial; Dreams; Understanding, of death
"Defence of Poetry, A" (Shelley), 51
Deification, 156, 165
DeLoache, Judy, 105
Dementia, 43, 49–50, 127
Denial of Death, The (Becker), 202
Denisovans, 52
Dennett, Daniel, 214
Depersonalization, 42–43, 48
Descartes, 42
Descent of Man, The (Darwin), 5
Diffusion tensor imaging (DTI), 13–14
Dinosaurs, 24
DNA, 11; of *Homo sapiens,* 72–73
Dobzhansky, Theodosius, 77, 113, 213
Dogs, 143, 148
Dolmen, 186
Dolphins, 44, 49
Domesticated Animals from Early Times (Clutton-Brock), 144
Domestication, 133; animal, 143–44; plant, 20, 140–42
Donald, Merlin, 40
Dostoevsky, Fyodor, 221
Dreams, 205; HRAF on, 119, 229–38; religion related to, 117–19
DTI. *See* Diffusion tensor imaging
Dumuzi, 168–69
Dunbar, Robin, 78–79, 81, 97, 201–2; on social brain hypothesis, 35–36
Durkheim, Émile, 123, 195–96, 206–7

Early *Homo sapiens. See* Neandertals
Eastern Apache Indians, 230–31
Eccles, John, 14, 77, 90, 106
Economics, 172–73

Edgar, Blake, 95
Egypt, 155, 196; death and, 178–82; grave goods in, 180–82; pyramids in, 179, 194–95; temple of, 177; writing of, 176–77
Egyptian gods, 177–78, 182
Einstein, Albert, 14, 34
Elementary Forms of the Religious Life, The (Durkheim), 207
Elephants, 44, 49, 115, 225–26
Eliot, T. S., 87, 107, 114
Empathy, 55–56, 60–61
End of Faith, The (Harris), 204
Engineering, 183
England: Silbury Hill, xiii–xiv, 188; Stonehenge, 134, 189, 190, 195, 222–23. *See also* Avebury
Enki, 166–68, 194, 208
Epic of Gilgamesh, The, 169–70
Epilepsy, 215
Episodic memory. *See* Autobiographical memory
Eskimos, Chugach, 139
Ethics, 169
Euhemeros of Macedonia, 156
Evidence: children as, 14–16; DTI in, 13–14; fMRI in, 13; hominin skulls as, 9–11; postmortem brains as, 12–13; tools as, 10; white matter connecting tracts in, 13–14, 15
Evolution, 35, 59, 242n18, 245n1; of human brain, 225–28; terminology about, xiv–xv. *See also* Parallel evolution; specific topics
Evolutionary theory, 3–6
Evolution of the Brain: Creation of the Self (Eccles), 90
Evolving God (King), 207
Executive functions, 162–63

Faces in the Clouds: A New Theory of Religion (S. Guthrie), 213–14
Fagan, Brian, 53, 55, 89
Faith Instinct, The (Wade), 207–8
Fear, 113–14
Feasting, 135–36, 143
Feder, Kenneth, 28, 38
Fertile Crescent, 140–44
Fertility, 160, 167–68
Figurines, 102, 184, 192–93; high gods related to, 159–60; Venus, 93, 97, 100
Fire, 39–40
First-order theory of mind, 56, 75
Fishing, 89, 102
Flechsig, Paul Emil, 7, 32, 163, 227, 276*n*5
fMRI. *See* Functional MRI
Food, 169, 180–81; cannibalism, 115; cereals, 140–42; cooking, 39; feasting, 135–36, 143; meat eating, 27. *See also* Domestication
Forebrain, 24, 242*n*1
Foresight, 259*n*35
Four Quartets (Eliot), 87, 107
France, 96–97, 100
France, Anatole, 30
Frazer, James, 222
Freud, Sigmund, 211
Frith, Chris, 56
Fromm, Erich, 116
Frontal cortex (BA 8), 61, 63, 64
Frontal lobe, 33, 127, 216; of *Homo habilis*, 30–31; theory of mind and, 63–64
Frontal pole (BA 10), 63, 64, 84, 85, 225; autobiographical memory and, 126, 127; intelligence and, 31–33, 32
Frontotemporal dementia, 49–50, 127
Functional MRI (fMRI), 13
Future, 106–8, 128–29

Gage, Phineas, 60–61
Gallup, Gordon, 42–43
Gargas cave art, 100, 102
Genetic theories, 217–18
Geography, 51–52
Geometric figures, 100–1
Germany, 97
Gervais, Will, 66

Ghosts, 230–31, 234–35
Gilgamesh, 169–70, 269*n*11
Gillings, Mark, 188–90
Gimbutas, Marija, 184
Glial cells, 6, 225–26, 276*n*1
Goats, 143–44
Göbekli Tepe, Turkey, 137–40, 143, 158; carved human figures in, 134–35; as ceremonial center, 135–36
God contests, 220–21, 275*n*38
God Delusion, The (Dawkins), 220
God Gene, The (Hamer), 217–18
"God gene," 217–19
Gods, xiii, 6, 23, 191, 240*n*3; adaptionists on, 218–20; anthropomorphism of, 83, 197; as by-products, 220–21; governments and, 170–72, 195–96, 200, 204–5, 205; introspective self and, 82–83; monotheism, 2, 196, 201; Neandertals and, 67; need for, 221–22; pattern-seeking theories and, 83; second-order theory of mind and, 82; from spirits, 156–58, 165; term use of, xv, 123; writing and, 160–61. *See also* specific topics
God's Brain (Tiger and McGuire), 212
Gold, 184–85
Goodall, Jane, 115
Gorillas, 43, 58
Gould, Stephen Jan, 24, 37
Governments, 4, 198; gods and, 170–72, 195–96, 200, 204–5, 205. *See also* Mesopotamian gods
Grave goods, 102–3, 148, 189, 194, 229; afterlife and, 125, 152; autobiographical memory and, 124–25; in China, 191–92; definition of, 94–95; in Egypt, 180–82; food as, 169, 180–81; of modern *Homo sapiens,* 92–95, 102–3; in Pakistan, 183–84; of royalty, 171–72; from Southeastern Europe, 184–85; Tylor on, 124–25; wealth in, 92–93; in Western Europe, 185–86
Great Plains, 230
Great Spirit, 230, 232, 240*n*3
Greeks, 156, 176, 178, 197, 200
Grids of connectivity, 8–9
Guthrie, Dale, 88, 124, 143, 161
Guthrie, Stewart, 212–14
Gyri, 10, 227, 277*n*6
Gyrification, 227–28, 277*n*6

Hallowell, A. Irving, 118
Hallucinations, 122, 218, 236
Hamer, Dean, 217–19
Hamilton, Edith, 178
Hamlet (fictitious character), 113
Handprints, 100, 121
Harappans, 182–84
Harris, Sam, 204, 221
Hawkes, Jacquetta, 150, 153
Hayden, Brian, 122, 142–43
Herodotus, 178, 201
Hick, John, 199
High gods, xv, 20, 138; population and, 158–59, 166; statues related to, 159–60
Hindbrain, 242*n*1
Hinde, Robert, 211–13
Hinduism, 196
Hippocampus, 126, 127–28, 216
History of God, A (Armstrong), 199
Hittites, 196
Hobbes, Thomas, 116
Hodder, Ian, 153, 160
Homer, 2
Hominids, xiv–xv
Hominins, xiv–xv, 9–11, 67; brain size of, 25–26; cognitive skill of, 3–4; death and, 111–12; Neandertals and, 69, 87
Homo erectus, 50–51, 255*n*16; brain size of, 38, 40, 45–46; fire and, 39–40; *Homo habilis* compared to, 37–38; migration of, 39–40; modern human and, 46–48, 47; self-awareness of, 42, 45–46, 129; tools of, 38–40, 245*n*3
Homo habilis, 10, 129, 245*n*3; brain size of, 26–27, 29–30, 37–38; *Homo erectus*

compared to, 37–38; parietal lobe of, 30–31; tools of, 27–28, 37
Homo sapiens, xiv–xv; autobiographical memory of, 3–4, 16, 19; brain size of, 29, 84; DNA of, 72–73; human brain of, 29, 84–86; introspective self of, 84–86, 85, 129; migration of, 11–12, 71–73, 87; self-ornamentation of, 10, 70–71, 74; in Siberia, 73. *See also* Modern *Homo sapiens*
Horses, 98–99
Houses, 136–37, 149, 152
How We Believe (Shermer), 214
HRAF. *See* Human Relations Area Files
Human-animal composites, 100
Human brain, 5, 204, 240*n*9, 241*n*16, 243*n*12; after death, 9; chimpanzee brains compared to, 12–13, 30, 33, 163, 226–27; conscription within, 35; description of, 6–7, 7, 8; divisions of, 7, 7, 8; evolution of, 225–28; growth of, 205–6, 245*n*20; gyrification in, 227–28, 277*n*6; gyri of, 10, 227, 277*n*6; of *Homo sapiens,* 29, 84–86; language related to, 9, 80–81, 241*n*10; maturation within, 12; of modern *Homo sapiens,* 125–29, 126; myelin in, 226–27; natural selection and, 16–17; of Neandertals, 84–86, 85, 129; network of, 8–9; neurons of, 6, 24, 30, 33, 48–50, 64–65, 225–26, 240*n*8, 241*n*12; occipital area of, 7, 7, 10, 14, 30; opportunistic evolution and, 35; social brain hypothesis related to, 35–36; terminal zones of, 7, 32, 163, 227; tissues and, 12–13, 241*n*12. *See also* Brain size; White matter connecting tracts
Human Connectome Project, 241*n*16
Human figures, 134–36, 138; in cave art, 99–100; statues, 151, 159–60, 172, 181. *See also* Figurines
Human Relations Area Files (HRAF), 119, 236–38; on Indians, 229–35
Human revolution, 101; autobiographical memory and, 120–21
Human sacrifices, 171–72
Humphrey, Nicholas, 28–29, 56, 76
Hunter-gatherers, 110, 119, 124, 134, 158; ancestor worship by, 138–40. *See also* Human Relations Area Files
Hunting, 38–39, 53, 70, 134; with autobiographical memory, 108–10

Ikram, Salma, 178
Immortality, 170, 204, 205
Indians: Bella Coola, 233; Blackfoot, 139; Canela, 235; Chipewyan, 234; Comanche, 230; Cree, 232–33; Creek, 229; Eastern Apache, 230–31; Inuit, 234; Mataco, 235; Nootkan, 233; Ojibwa, 118, 139, 232; Pawnee, 119; Pomo, 231; Stoney, 232; Tlingit, 231–32; Ute, 230; Yoruk, 231
Indonesia, 96
Inferior parietal cortex, 126
Inferior parietal lobe (BA 39 and 40), 46, 47, 48, 228, 248*n*23; intelligence and, 31–32, 32, 34–35; theory of mind and, 63, 126, 126
In Gods We Trust (Atran), 220
Insula, 84, 86; self-awareness and, 46–47, 47; theory of mind and, 60, 62–64. *See also* Anterior insula
Intelligence, 46; BA 9 and 46 related to, 31–32, 32; BA 24 and 32 and, 31–32, 32; BA 39 and 40 and, 31–32, 32, 34–35; corpus callosum and, 34
Intentional burials: debates about, 94–95; of Neandertals, 54–55. *See also* Grave goods
Intentional cultivation, 141
Introspective self, 205, 255*n*16; animals without, 77; cognition of, 76–77; gods and, 82–83; of *Homo sapiens,* 84–86, 85, 129; language and, 78–82; neuroimaging studies on, 84–86, 85; white matter connecting tracts and, 86

Inuit Indians, 234
Israel, 151–52
Italy, 94

Jacobsen, Thorkild, 167–68, 170–71
James, William, 104–5, 123, 199
Jaspers, Karl, 199
Jaynes, Julian, 218
Johnson, Dominic, 65
Jordan, 151
Joseph, Rhawn, 216
Judaism, 201
Jung, Carl, 1

Ka, 179, 181
Kalahari San, 236
Kennett, Douglas, 141
Kfar Hahoresh, Israel, 151–52
King, Barbara, 207
Kings, 170–72, 205; Tutankhamun, 180–81, 196
1 Kings 18:20–40, 275*n*38
Klein, Richard, 95
Koryak, 238
Kramer, Samuel, 173

Lamps, 89, 103
Land ownership, 147
Language: autobiographical memory and, 110; cognition before, 79–80; human brain related to, 9, 80–81, 241*n*10; introspective self and, 78–82; linguistics and, 81; of Mesopotamia, 166, 169, 269*n*8; origin of, 78–79; sign, 44; social grooming and, 78–79
Langworthy, Orthello, 226
Lascaux cave art, 96, 99–100, 103, 123
Lateral prefrontal cortex (BA 9 and 46), 84, 85; of children, 163; intelligence and, 31–32, 32; planning related to, 162, 162–64
Leary, Mark, 20, 80
Letter from a Region of My Mind (Baldwin), 133
Letter to a Christian Nation (Harris), 221
Lewis-Williams, David, 110, 122
Lice, 71
Linden, David, 212
Linguistics, 81
Lobotomies, 61
Longshan culture, 190–92
Long-term memory, 105. *See also* Autobiographical memory
"Lucy," 26
Luther, Martin, 199

Macaques, 24–25
Magnetic resonance imaging (MRI), 13
Major religions: adherents of, 200; birth of, 198–99; borrowing by, 200–1; death and, 199; government and, 200; support from, 199–200, 205
Malaysia, 237–38
Malraux, André, 165, 187
Mammalian brains, 18, 23–24
Manitous (spirits), 232–33
Man's Fate (Malraux), 165
Manus Islanders, 237
Maoris, 118
Marshack, Alexander, 90
Marsupial mammals, 18
Masks, 151, 157–58
Mataco Indians, 235
Mathnawī (Rūmi), 68
Mbuti Pygmies, 236
McBrearty, Sally, 101
McGovern, Patrick, 141–42, 146, 154
McGuire, Michael, 212
McNamara, Patrick, 1, 118–19, 216–17, 219
Meat eating, 27
Medial prefrontal cortex, 46, 63–64, 127, 162
Meditation, 216
Mellaart, James, 159–60
Mellars, Paul, 70, 96–97
Memory, 259*n*35; semantic, 105, 107–9. *See also* Autobiographical memory
Memory devices, 90–91, 102

Mesoamerica, 146
Mesopotamia: Assyria and, 196; language of, 166, 169, 269*n*8; population of, 166; society of, 166; writing of, 166, 169–70, 175
Mesopotamian gods, 166; anthropomorphism of, 172, 175–76; death and, 169–71; Egyptian gods compared to, 177; fertility and, 167–68; kings related to, 170–72; seasons related to, 169; temples for, 167, 172–74; war related to, 174–75
Mesulam, Marcel, 20
Metallurgy, 184–85
Midbrain, 24, 242*n*1
Middle East, 69
Migration: of animals, 109; of *Homo erectus,* 39–40; of *Homo sapiens,* 11–12, 71–73, 87
Minds, 28–29, 129; bicameral, 218. *See also* Theory of mind
Mirror neurons, 64–65
Mirror self-recognition, 41, 43–45
Mithen, Steven, 27–28, 82, 90, 125
Modern *Homo sapiens,* 87; arts of, 95–101, 102–3; autobiographical memory of, 104–9, 125–29, 126; fitted clothing of, 71, 89; grave goods of, 92–95, 102–3; human brain of, 125–29, 126; memory devices of, 90–91, 102; self-ornamentation of, 91–92, 102–3; tools of, 88–89; trade networks of, 91–92; weapons of, 88–90
Modern human, 46–48, 47
Modular Brain, The (Restak), 60
Modules, 8–9
Monkeys, 24–25; brains of, 18–19; chimpanzees compared to, 43–44, 49; tools of, 27
Monotheism, 2, 196, 201
Montaign, Michel de, 2
Montesquieu (baron), 176
Mormonism, 211, 222
Moseley, Michael, 193–94, 197–98
Mother Goddess, 159–60
Mounds: burial, 223; platform, xiv, 192–93, 195
MRI. See Magnetic resonance imaging
Mummification, 179–80; in Peru, 193–94
Murdoch, George, 158
Musical instruments, 97, 102, 154, 193
Myelin, 226–27

Nabokov, Vladimir, 114
Nakoda (Stoney) Indians, 232
Natural phenomena, 83
Natural selection, 5–6; human brain and, 16–17
Neandertals (Archaic *Homo Sapiens*): art by, 95; autobiographical memory and, 108–9; brain size of, 52, 68–69, 84; burials of, 54–55, 92, 94–95; culture of, 53–54; death and, 112; description of, 52–53; geography of, 51–52; gods and, 67; Hominins and, 69, 87; human brain of, 84–86, 85, 129; intentional burials of, 54–55; tools of, 53, 249*n*5
Needles, 88–89, 102
Neocortex, 24
Neolithic era, 160, 265*n*24
Neuroimaging studies, 31, 48; on introspective self, 84–86, 85
Neurological theories, 215–17
Neurons, 6, 24, 30, 33, 240*n*8, 241*n*12; in brain evolution, 225–26; self-awareness and, 48–50; theory of mind and, 64–65
Neuropsychological Bases of God Beliefs (Persinger), 215
Newberg, Andrew, 216
New World monkeys, 18–19
New Zealand, 118
Nootkan Indians, 233
Norenzayan, Ara, 65
Northern Canada, 234

Occipital area, 7, 7, 10, 14, 30
Ochre, 53–54, 70–71
Oedipal complex, 211

Ojibwa Indians, 118, 139, 232
Old World monkeys, 18–19
Olszewski, Deborah, 145
On the Origin of Species (Darwin), 110–11
Ontogony and Phylogeny (Gould), 37
Opportunistic evolution, 35
Orbital frontal cortex (BA 47), 85, 126, 127, 216
Orbitofrontal region (BA 47), 84, 85, 126, 127
Origin of Consciousness and the Breakdown of the Bicameral Mind, The (Jaynes), 218
Osiris, 177, 179–80, 182
Other-than-human-persons, 232
Ozymandias, 223

Pakistan, 182; Ahmadiyya in, 222; Aryans in, 196; engineering in, 183; grave goods in, 183–84
Paleolithic period, 89–91, 133, 239*n*2. *See also* specific topics
Papua New Guinea, 146, 156–58
Parahippocampal gyrus (BA 35, 36), 126, 127–28
Parallel evolution, 17, 206, 242*n*18; of agriculture, 145–46; chimpanzees and, 25; examples of, 18–20; metalurgy, 185
Parietal lobe, 33, 215–16; of *Homo habilis,* 30–31
Parrinder, Geoffrey, 139–40
Passingham, Richard, 79–80
Pattern-seeking theories, 83, 213–14
Patterns in Prehistory (Olszewski and Wenke), 145
Pawnee Indians, 119
Pearson, Mike Parker, 23, 113, 147–50
Pech Merle cave art, 98–99, 103
Persinger, Michael, 215
Peru, xiii–xiv, 146, 154; mummification in, 193–94; pyramids in, 192–93; society of, 192; temple in, 197–98
Petronius, Gaius, 115–16
Phenomenon of Man, The (Teilhard de Chardin), 255*n*16
Piaget, Jean, 14–15
Picasso, Pablo, 99
Pigs (boars), 144
Pinker, Steven, 80
Placental mammals, 18
Planning, 28; BA 9 and 46 related to, 162, 162–64
Plant domestication, 20; cereals in, 140–42; intentional cultivation in, 141
Plastered skulls, 149–51, 159, 266*n*37
Platform mounds, xiv, 192–93, 195
Politics, 4
Pollard, Joshua, 188–90
Polytheism, 2, 240*n*3
Pomo Indians, 231
Population, 198, 205; high gods and, 158–59, 166
Posterior cingulate (BA 23), 85, 85, 128
Posterior superior temporal area (BA 22), 62, 63, 126, 126
Postmortem brains, 12–13
Potassium-argon dating, 11
Potlatch ceremonies, 139
Pottery, 145–46, 184, 187, 191–92, 265*n*24
Povinelli, Daniel, 15–16, 56–57, 104
Power, 190; from ancestor worship, 155–56
Precision grip, 18–19
Prefrontal cortex, 277*n*6; agriculture and, 163–64; medial, 46, 63–64, 127, 162; theory of mind and, 60–61. *See also* Lateral prefrontal cortex
Prehistory of the Mind (Mithen), 90
Premotor cortex (BA 6), 31–33, 32
Preuss, Todd, 33, 163
Primates: baboons, 24–25, 42, 58, 115; brains of, 33; gorillas, 43, 58; monkeys, 18–19, 24–25, 27, 43–44, 49. *See also* Chimpanzees
Primitive Culture (Tylor), 110–11, 125
Prince, Christopher, 56–57
Principles of Brain Evolution (Striedter), 24, 245*n*20
Prosocial theories, 208–10, 273*n*9

Proust, Marcel, 105–6
Psychological theories, 211
Pullum, Geoffrey, 80
Pyramids, xiii–xiv, 187; in Avebury, 188; in Egypt, 179, 194–95; in Peru, 192–93

Queens, 107, 172

Radioactive thorium dating, 11
Radiocarbon dating, 11
Ramachandran, Vilayanur, 215
Rapid-eye-movement (REM) sleep, 117
Religion, 222–23; cave art related to, 121–24; definitions of, 123, 207; dreams related to, 117–19; politics and, 4; term use of, xv; understanding of death and, 115–16. *See also* Major religions
Religion Explained (Boyer), 214, 220
Religious symbols, 77, 83
REM. *See* Rapid-eye-movement (REM) sleep
Remembrance of Things Past (Proust), 105–6
Renfrew, Colin, 185
Restak, Richard, 60
Right temporo-parietal junction (RTPJ), 62–63, 63
Roots of Civilization, The (Marshack), 90
Rose, Michael, 29–30
Rosenberg, Michael, 136
Rousseau, Jean-Jacques, 2
Roux, George, 172, 176
Royalty, 107, 170–72, 205
RTPJ. *See* Right temporo-parietal junction
Rūmi, 68
Russia, xiv, 73, 186, 238

Sacrifices, human, 171–72
Sahelanthropus tchadensis, 25
Sakhalin Island, Russia, 238
Sally-Anne test, 56–57, 75
Samoa, 120
Saviors, 201, 222
Schizophrenia, 42–43
Schmidt, Klaus, 135, 137
Schoenemann, Thomas, 79, 241*n*12
Scotland, 187–88
Sculptures, 97; statues, 151, 159–60, 172, 181. *See also* Figurines
Seasons, 169
Second-order theory of mind, 75–76, 82
Self-awareness, 205, 247*n*17; age related to, 44–45; of children, 40–41; of chimpanzees, 43–44; definition of, 41–42; empathy related to, 55–56; of *Homo erectus*, 42, 45–46, 129; insula and, 46–47, 47; neurons and, 48–50; VENs related to, 48–50
Self-ornamentation, 69, 76; of *Homo sapiens*, 10, 70–71, 74; of modern *Homo sapiens*, 91–92, 102–3
Self-recognition, mirror, 41, 43–45
Semantic memory, 105, 109; brain abnormalities and, 107–8
Sewing, 88–89, 102
Shabtis, 181
Shakespeare, William, 203
Shamans, 122, 235, 238
Shamans, Sorcerers, and Saints (Hayden), 122
Shamans of Prehistory, The (Clottes and Lewis-Williams), 122
Shang dynasty, 197
Shariff, Azim, 158–59
Sheep, 143–44
Shell beads, 70–71
Shelley, Percy Bysshe, 51
Shermer, Michael, 214
Shiva, 183
Short-term memory (working memory), 105
Shrines, xiii–xiv, 152–53
Shryock, Andrew, 38
Siberia, 73, 119–20
Sign language, 44
Silbury Hill, England, xiii–xiv, 188
Skull cults, 151–53; display of, 149–50

Skull house, 136–37
Skulls, plastered, 149–51, 159, 266*n*37
SLF. *See* Superior longitudinal fasciculus
Smail, Daniel, 17, 38
Smith, Joseph, 211
Social brain hypothesis, 35–36
Social cognition, 84–85
Social grooming, 78–79
Social theories, 206–8
Solstice, 189, 222–23
Sorcerers, 100, 122–23, 262*n*65
Soul traveling, 238
South Africa, 69, 70–71
Southeastern Europe, 184–85, 197
Spain, 96–97
Spear throwers, 89, 102–3
Spencer, Herbert, 156
Spirits, 230, 240*n*3; Ba, 179; gods from, 156–58, 165; manitous as, 232–33
Stanley Medical Research Institute, 243*n*12
Statues, 151, 172, 181; high gods related to, 159–60
Stonehenge, 134, 190, 195; solstice at, 189, 222–23
Stones, 186–87
Stoney (Nakoda) Indians, 232
Striedter, Georg, 24, 245*n*20
Stringer, Christopher, 61
Suddendorf, Thomas, 110, 259*n*35
Suffering, 6
Sumer, 269*n*8, 269*n*11
Superior longitudinal fasciculus (SLF), 34, 46, 163
Superior parietal (BA 7), 31–32, 32, 34, 126, 128
Survival, 219–20. *See also* Death
Swanson, Guy, 158
Swift, Jonathan, 30
Synapses, 277*n*6

Tallis, Raymond, 45, 112
Taste, 106
Tattersall, Ian, 45, 55, 95, 125
Teilhard de Chardin, Pierre, 77, 82, 255*n*16
Temples: children and, 173; of Egypt, 177; for Mesopotamian gods, 167, 172–74; in Peru, 197–98
Temporal lobe, 215
Temporal pole (BA 38), 85, 85–86, 256*n*30
Temporo-parietal junction (TPJ), 62–63, 63, 85
Terminal zones, 7, 32, 163, 227
Terror Management Theory, 116–17
Theories: comfort, 211–13; evolutionary, 3–6; genetic, 217–18; neurological, 215–17; pattern-seeking, 83, 213–14; prosocial, 208–10, 273*n*9; psychological, 211; social, 206–8
Theory of mind, 250*n*12; animals and, 57–59; BA 24 and 32 and, 60, 63, 64, 84, 85, 86, 126, 126; belief in gods and, 65–67; children and, 56–57, 74–75, 82; empathy and, 60–61; evolution and, 59; first-order, 56, 75; impairment of, 59–61; insula and, 60, 62–64; mirror neurons related to, 64–65; neurons and, 64–65; prefrontal cortex and, 60–61; second-order, 75–76, 82; TPJ in, 62–63, 63; uncinate fasciculus and, 62
Through the Looking Glass (Carroll), 107
Tiger, Lionel, 212
Tillich, Paul, 113
Time. *See* Autobiographical memory
Tissues, 12–13, 241*n*12
Tlingit Indians, 231–32
Tobias, Philip, 29–30, 35
Tools, 10, 48, 69; bone, 70, 88, 102; of *Homo erectus,* 38–40, 245*n*3; of *Homo habilis,* 27–28, 37; of modern *Homo sapiens,* 88–89; of Neandertals, 53, 249*n*5
Totem poles, 135–36
Totems, 121–23
Toynbee, Arthur, 196
TPJ. *See* Temporo-parietal junction
Trade networks, 91–92

Trigger, Bruce, 182
Tulving, Endel, 259*n*34
Turgenev, Ivan, 30
Turkey, 136, 152–53, 159–60. *See also* Göbekli Tepe
Tutankhamun (king), 180–81, 196
Tylor, Edward B., 114, 117–18, 123, 138; on ancestor worship, 140, 156, 204; Darwin and, 110–11, 206; on grave goods, 124–25

Uncinate fasciculus, 14, 15, 46–47; autobiographical memory and, 128; theory of mind and, 62
Understanding, of death: animism and, 114; autobiographical memory and, 110–13; children related to, 114–15; religion and, 115–16; Terror Management Theory and, 116–17
UNESCO World Heritage site, 193
Upper Paleolithic period, 90, 110, 239*n*2
Uranium dating, 11
Urgesi, Cosimo, 215–16
Utah, 230
Ute Indians, 230
Utnapishtim, 170

Veddahs, 138–39, 236
VENs. *See* von Economo neurons
Venus figurines, 93, 97, 100
Villages, 148
Virgin birth, 201, 272*n*76
Visual arts, 19; autobiographical memory and, 120–22. *See also* Cave art
von Economo, Constantin, 49
von Economo neurons (VENs), 48–50

Wade, Nicholas, 207–8, 219
WAIS. *See* Wechsler Adult Intelligence Scale
War, 174–75, 190
Warnings, 234
Water crossing, 73
Wealth, 171–72
Weapons: bow and arrow, 89, 103; of modern *Homo sapiens,* 88–90; spear throwers, 89, 102–3
Wechsler Adult Intelligence Scale (WAIS), 31–32
Weil, Eric, 199
Wenke, Robert, 145
Wernicke's area, 9, 62, 81
Western Europe: Brodgar in, 187–88; cairns in, 186–87; grave goods in, 185–86. *See also* England
Whales, 30, 44, 49, 225–26
What Mad Pursuit (Crick), 1
Wheeler, Mortimer, 183
White, Randall, 91, 101
White matter connecting tracts, 34, 46–47, 226; in evidence, 13–14, 15; introspective self and, 86; planning and, 163–64
White people, 157
Why Gods Persist (Hinde), 211–12
Willis, Thomas, xiii
Wilson, David Sloan, 207–8
Wine, 141–42, 154
Women: figurines, 93, 97, 100, 159–60; sewing by, 88–89, 102; virgin birth, 201, 272*n*76
Working memory. *See* Short-term memory
Wren, Christopher, 239*n*1
Writing, 191; afterworld in, 181–82; of Egypt, 176–77; gods and, 160–61; of Mesopotamia, 166, 169–70, 175

Xenophanes, 176

Yeats, William Butler, 116
Yoruk Indians, 231
Younger Dryas, 263*n*2

Zilles, Karl, 48, 227
Zimmer, Carl, 72–74, 239*n*1
Zoroaster, 201, 272*n*76
Zoroastrianism, 201, 272*n*76
Zulu, 118